Microelectronic Systems

Microelectronic Systems

Microelectronic Systems

R. Vears

Newnes
an imprint of Butterworth-Heinemann
Linacre House, Jordan Hill, Oxford OX2 8DP
A division of Reed Educational and Professional Publishing Ltd

A member of the Reed Elsevier plc group

OXFORD   BOSTON   JOHANNESBURG
MELBOURNE   NEW DELHI   SINGAPORE

First published 1996

British Library Cataloguing in Publication Data
Vears, R. E.
   Microelectronic systems
   1 Microelectronics 2 Electronic systems 3 Microelectronics – Problems, exercises, etc.
   4 Electronic systems – Problems, exercises, etc.
   I Title
   621.3'81'011

ISBN 0 7506 2819 7

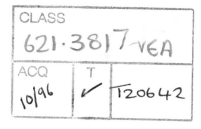
Printed and bound in Great Britain by The Bath Press, Bath

Contents

Preface

This book has been written to provide coverage of microelectronic systems (Unit 19) for the Advanced GNVQ in Engineering, although it may regarded as a general textbook for a much wider range of studies in microelectronic systems.

The aim of this book is to provide a foundation in hardware, software and interfacing techniques for microelectronic systems, and each topic is presented in a way that assumes little prior knowledge of the subject.

Three different types of microprocessor are used in the text as examples. First the well-established Zilog Z80 (or later derivatives) is presented as an example of a general purpose 8-bit microprocessor that offers students an ideal introduction to the principles of microelectronic systems. Next the Intel 8086 (80286, 80386, 80486 etc.) microprocessor is included because of its wide use in personal desktop computers, thus allowing students to gain experience of 16/32-bit microprocessor systems. Finally the Intel 8031/32 (or 8051/52) range is considered to provide students with an introduction to the somewhat different architecture and instruction set associated with single chip microcontrollers.

Structured programming is encouraged with the aim of applying engineering principles to software development, rather than resorting to haphazard inspirational techniques which often lead students to believe that programming is too difficult a task for them. Examples in the text use structured programming techniques and include pseudo-code, making it a relatively simple matter to write equivalent assembly language programs for most common types of microprocessors.

The principles of interfacing are introduced in general terms. However, the wide range of interfacing techniques used, and numerous specialized devices available, prevent comprehensive coverage of all aspects of interfacing in a book of this type. Therefore only the more common interfacing problems are dealt with in any detail.

Self-test questions and answers are included in the text to enable students to check their knowledge, and activities are suggested which should help provide much of the documentary evidence required for assessment of the subject.

Finally I wish to express appreciation to all those who have given me encouragement and assistance, the publishers for their advice, and special thanks to my wife Rosemary and our family for their patience during the preparation of this book.

Ron Vears

1 Number systems and computer arithmetic

Summary

This element deals with the theory of number systems and introduces binary and hexadecimal number systems which are essential for an understanding of the operation of microprocessors. Conversions between the different number systems are dealt with, as are the concepts of addition, subtraction and complementation for representing negative numbers in a microprocessor. A large number of worked examples are included in the text with further problems as self-test exercises.

Number systems

Denary system

In everyday situations, counting takes place using a **denary** (base ten) or **decimal** system. Its main justification for use is often quoted as being that human beings have ten fingers/thumbs with which to count. The characteristics of a denary system are:

1   a set of ten distinct counting digits (0, 1, 2, 3, 4, 5, 6, 7, 8 and 9), and
2   a place value (or weight) for each digit, organized in ascending powers of ten starting from the right.

Thus, for example, the denary number 2658_{10} may be considered as follows:

$$2658_{10} = (2 \times 10^3) + (6 \times 10^2) + (5 \times 10^1) + (8 \times 10^0)$$

A denary system is unsuitable for use in electronic systems, since practical limitations imposed by electronic devices allow only two conditions to be consistently predictable. These conditions are obtained when a chosen electronic device is made to act as a switch, and its two states are on and off, represented by digits 0 and 1. Therefore there are insufficient counting digits for the denary system to be used.

Binary system

A binary (base two) number system is suitable for use in electronic equipment. Binary digits 0 and 1 are called **bits** and a digital electronic system performs its tasks by processing information that is represented by groups of bits.

The characteristics of the binary system are:

1 a set of two distinct counting digits (0 and 1), and
2 a place value (or weight) for each digit, organized in ascending powers of two starting from the right.

Thus a '1' in a particular position in a binary number contributes its place value towards the total, but a '0' contributes nothing. Therefore the denary equivalent of a binary number may be obtained by adding together all of the place values where a '1' occurs in the binary number.

Example 1

Convert the binary number 1101_2 to denary.

$$1101_2 = (1 \times 2^3) + (1 \times 2^2) + (0 \times 2^1) + (1 \times 2^0)$$

$$= (1 \times 8) + (1 \times 4) + (0 \times 2) + (1 \times 1)$$

$$= 8 + 4 + 0 + 1$$

$$= \mathbf{13_{10}}$$

Thus the denary equivalent of 1101_2 is 13_{10}.

Example 2

Convert the binary number 10110111_2 to denary.

$$10110111_2 = (1 \times 2^7) + (0 \times 2^6) + (1 \times 2^5) + (1 \times 2^4) +$$
$$(0 \times 2^3) + (1 \times 2^2) + (1 \times 2^1) + (1 \times 2^0)$$

$$= (1 \times 128) + (0 \times 64) + (1 \times 32) + (1 \times 16) +$$
$$(0 \times 8) + (1 \times 4) + (1 \times 2) + (1 \times 1)$$

$$= 128 + 0 + 32 + 16 + 0 + 4 + 2 + 1$$

$$= \mathbf{183_{10}}$$

Thus the denary equivalent of 10110111_2 is 183_{10}.

In order to reverse this process, that is, convert a denary number into its binary equivalent, several methods are available. One method commonly adopted is the two's method in which a denary number is repeatedly divided by two until a quotient of zero is obtained. After each division, the remainder (which can only be 0 or 1) is noted and used to form the binary equivalent.

Example 3

Convert the denary number 25_{10} to binary.

25/2	=	12	remainder 1	least significant digit (LSD)
12/2	=	6	remainder 0	
6/2	=	3	remainder 0	
3/2	=	1	remainder 1	
1/2	=	0	remainder 1	most significant digit (MSD)

Thus the binary equivalent of 25_{10} is 11001_2.

Example 4

Convert the denary number 147_{10} to binary.

147/2	=	73	remainder 1	least significant digit (LSD)
73/2	=	36	remainder 1	
36/2	=	18	remainder 0	
18/2	=	9	remainder 0	
9/2	=	4	remainder 1	
4/2	=	2	remainder 0	
2/2	=	1	remainder 0	
1/2	=	0	remainder 1	most significant digit (MSD)

Thus the binary equivalent of 147_{10} is 10010011_2.

Although binary numbers are essential for internal processing by a digital electronic system, they are rather tedious and error-prone when used by human operators. This is particularly the case when dealing with binary numbers in a microcomputer when there may be 32 or more bits in each number. Outside of a computer some other form of number system is required; however, the denary system with which we are familiar is not the most logical to use. The previous examples show that it is difficult to see any obvious relationship between binary and denary digits, caused by the fact that ten is not an exact power of two; therefore a hexadecimal number system is often preferred.

Test your knowledge

1 Convert the following binary numbers to denary:
   (a) 1110_2 (14_{10})
   (b) 11111_2 (31_{10})
   (c) 11110001_2 (241_{10})
   (d) 10111010_2 (186_{10})
   (e) 11001100_2 (204_{10})
   (f) 10001101_2 (141_{10})

2 Convert the following denary numbers to binary:
   (a) 18_{10} (10010_2)
   (b) 42_{10} (101010_2)
   (c) 176_{10} (10110000_2)
   (d) 98_{10} (1100010_2)
   (e) 225_{10} (11100001_2)
   (f) 138_{10} (10001010_2)

Hexadecimal system

A hexadecimal (base 16) or hex number system is often used to represent data outside of a digital electronic system. The characteristics of a hexadecimal number system are:

1 16 distinct counting digits (0, 1, 2, 3, 4, 5, 6, 7, 8, 9, A, B, C, D, E and F), and
2 a place value (or weight) for each digit, organized in ascending powers of 16 starting from the right.

Note that the additional six counting digits needed for this number system are provided by the first six letters of the alphabet (upper or lower case). Thus letters **A** to **F** correspond to denary numbers **10** to **15** (see Table 1.1).

Example 5

Convert the hexadecimal number $3B_{16}$ to denary.

$$3B_{16} \quad = \quad (3 \times 16^1) + (B \times 16^0)$$
$$= \quad (3 \times 16) + (11 \times 1)$$
$$= \quad 48 + 11$$
$$= \quad 59_{10}$$

Thus the denary equivalent of $3B_{16}$ is 59_{10}.

Table 1.1

Denary	Binary	Hexadecimal
0	0	0
1	1	1
2	10	2
3	11	3
4	100	4
5	101	5
6	110	6
7	111	7
8	1000	8
9	1001	9
10	1010	A
11	1011	B
12	1100	C
13	1101	D
14	1110	E
15	1111	F
16	10000	10
17	10001	11
18	10010	12
19	10011	13
20	10100	14

Example 6

Convert the hexadecimal number $2AF_{16}$ to denary.

$$2AF_{16} = (2 \times 16^2) + (A \times 16^1) + (F \times 16^0)$$
$$= (2 \times 256) + (10 \times 16) + (15 \times 1)$$
$$= 512 + 160 + 15$$
$$= 687_{10}$$

Thus the denary equivalent of $2AF_{16}$ is 687_{10}.

In order to reverse this process, that is, convert a denary number into its hexadecimal equivalent, repeated division by 16 may be used, similar to the method used for denary to binary conversion. In this case, a denary number is repeatedly divided by 16 until a quotient of zero is obtained. After each division, the remainder (expressed as a hexadecimal digit) is noted and used to form the hex equivalent.

Example 7

Convert the denary number 92_{10} to hexadecimal.

$$92/16 \quad = \quad 5 \qquad \text{remainder} \quad 12 \quad (= C)$$
$$5/16 \quad = \quad 0 \qquad \text{remainder} \quad 5$$

Thus the hex equivalent of 92_{10} is $5C_{16}$.

Example 8

Convert the denary number 6796_{10} to hexadecimal.

$$6796/16 \quad = \quad 424 \qquad \text{remainder} \quad C$$
$$424/16 \quad = \quad 26 \qquad \text{remainder} \quad 8$$
$$26/16 \quad = \quad 1 \qquad \text{remainder} \quad A$$
$$1/16 \quad = \quad 0 \qquad \text{remainder} \quad 1$$

Thus the hex equivalent of 6796_{10} is $1A8C_{16}$.

Conversion between binary and hex is much simpler, since each hex digit corresponds directly to a group of four binary digits. For example, consider the binary number 1110011110101001_2 which may be expressed in hex form as follows:

1 Group bits in fours from the right	1110	0111	1010	1001
2 Assign a hex digit to each group	E	7	A	9

Thus the hex equivalent of 1110011110101001_2 is $E7A9_{16}$.

Converting from hex into binary is simply the reverse of this process. For example, consider the hex number $A3FB_{16}$ which may be expressed in binary as follows:

1 Space out the hex digits	A	3	F	B
2 Convert each hex digit into binary	1010	0011	1111	1011

Thus the binary equivalent of $A3FB_{16}$ is 1010001111111011_2.

Test your knowledge

1 Convert the following hex numbers to denary:
  (a) $7A_{16}$ (122_{10})
  (b) $2F_{16}$ (47_{10})
  (c) $C9_{16}$ (201_{10})
  (d) BD_{16} (189_{10})
  (e) 98_{16} (152_{10})

2 Convert the following denary numbers to hex:

(a) 37_{10} (25_{16})
(b) 108_{10} $(6C_{16})$
(c) 161_{10} $(A1_{16})$
(d) 238_{10} (EE_{16})
(e) 54_{10} (36_{16})

3 Convert the following binary numbers to hex:

(a) 11010111_2 $(D7_{16})$
(b) 11101010_2 (EA_{16})
(c) 10001011_2 $(8B_{16})$
(d) 10100101_2 $(A5_{16})$
(e) 10011110_2 $(9E_{16})$

4 Convert the following hex numbers to binary:

(a) $1F_{16}$ (00011111_2)
(b) $A5_{16}$ (10100101_2)
(c) $6A_{16}$ (01101010_2)
(d) $2B_{16}$ (00101011_2)
(e) ED_{16} (11101101_2)

Computer arithmetic

Addition

The rules for the addition of any two numbers of the same base are as follows:

1 The equation for addition is $X + Y = Z$ where X is called the **augend**, Y is called the **addend** and Z is called the **sum**.
2 Digits in corresponding positions in each number are added, starting from the right.
3 A carry into the next most significant position occurs if the sum of two digits equals or exceeds the base used.

Binary addition

When adding binary numbers, a carry out is generated when a sum equals or exceeds two and this becomes a carry in for any subsequent addition. Binary addition rules are shown in Table 1.2.

Table 1.2

Augend	+	Addend	+	Carry in	=	Sum	,	Carry out
0		0		0		0		0
0		1		0		1		0
1		0		0		1		0
1		1		0		0		1
0		0		1		1		0
0		1		1		0		1
1		0		1		0		1
1		1		1		1		1

Example 9

Use the rules for binary addition to calculate the sum of $0110_2 + 0111_2$.

```
Augend              0    1    1    0    ⎫
Addend              0    1    1    1    ⎬  Added
                   ─────────────────    ⎭
Carry out ◄─ 0      1    1    0    0 ◄── Carry in

Sum                 1    1    0    1
                   ─────────────────
```

Thus the sum of **0110_2 + 0111_2** is **1101_2**.

In the absence of any following addition, the carry out becomes the most significant digit of the sum.

Example 10

Use the rules for binary addition to calculate the sum of 11101100_2 + 11111010_2.

```
Augend    1   1   1   0   1   1   0   0  ⎫
Addend    1   1   1   1   1   0   1   0  ⎬  Added
                                         ⎭
      ◄─1 1   1   1   1   0   0   0   0 ◄── Carry in
Carry out ───────────────────────────────

Sum  1    1   1   1   0   0   1   1   0
          ───────────────────────────────
```

In the absence of any following addition, the carry out becomes the most significant digit of the sum.

Thus the sum of **11101100_2 + 11111010_2** is **111100110_2**.

Hexadecimal addition

The rules for the addition of two hex numbers are similar to those for binary addition except that a carry out occurs when a sum equals or exceeds 16.

The most likely cause of problems when adding hex numbers is difficulty in determining a sum where mixtures of figures 0 to 9 and letters A to F are involved.

Example 11

Using the rules for hex addition, calculate the sum of $9A_{16} + B7_{16}$.

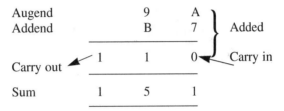

Augend		9	A
Addend		B	7
Carry out	1	1	0
Sum	1	5	1

Thus the sum of $9A_{16} + B7_{16}$ is 151_{16}.

Example 12

Using the rules for hex addition, calculate the sum of $89_{16} + A6_{16}$.

Augend		8	9
Addend		A	6
Carry out	1	0	0
Sum	1	2	F

Thus the sum of $89_{16} + A6_{16}$ is $12F_{16}$.

Subtraction

The rules for the subtraction of any two numbers of the same base are as follows:

1   The equation for subtraction is $X - Y = Z$ where X is called the **minuend**, Y is called the **subtrahend** and Z is called the **difference**.

2   Digits in the subtrahend are subtracted from digits in corresponding positions in the minuend, starting from the right.
3   If the difference obtained is less than zero, a *'borrow in'* from the next most significant position of the minuend is made (this may be achieved by adding 1 to the next most significant position of the subtrahend). The value of a 'borrow in' is equal to the base used.

Binary subtraction

The rules for binary subtraction are summarized in Table 1.3.

Table 1.3

Minuend	–	Subtrahend	=	Difference	,	Borrow
0		0		0		0
1		0		1		0
1		1		0		0
0		1		1		1

Example 13

Use the rules for binary subtraction to calculate $1000_2 - 0011_2$.

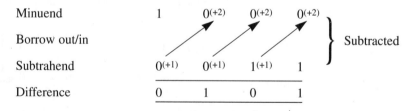

Therefore $\mathbf{1000_2 - 0011_2 = 0101_2}$.

Note that when necessary, a number equivalent to the base is borrowed (borrow in), and this is shown as (+2) in the minuend. To compensate for this borrow, 1 must be added to the subtrahend (borrow out) and this is shown as (+1) in the subtrahend (probably easier than trying to subtract 1 from the equivalent digit in the minuend, since this is often zero).

Example 14

Use the rules for binary subtraction to calculate $1110_2 - 1011_2$.

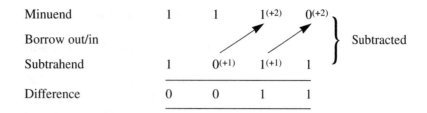

Minuend	1	1	$1^{(+2)}$	$0^{(+2)}$	
Borrow out/in					Subtracted
Subtrahend	1	$0^{(+1)}$	$1^{(+1)}$	1	
Difference	0	0	1	1	

Therefore $1110_2 - 1011_2 = 0011_2$.

Hex subtraction

This is similar to binary subtraction; however, problems are likely to be experienced due to the mixture of figures 0 to 9 and letters A to F in the hex numbering system.

Example 15

Using the rules for hex subtraction, calculate $A7_{16} - 6E_{16}$.

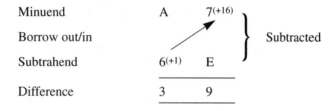

Minuend	A	$7^{(+16)}$	
Borrow out/in			Subtracted
Subtrahend	$6^{(+1)}$	E	
Difference	3	9	

Since E is larger than 7, a *borrow in* is required, shown as (+16). This means that E must be subtracted from 23 (i.e. 23 – 14 = 9). To compensate for this, 1 must be added to the next most significant position of the subtrahend, shown as (+1). Therefore 7 must be subtracted from A (i.e. 10 – 7 = 3).

Therefore $A7_{16} - 6E_{16} = 39_{16}$.

Example 16

Using the rules for hex subtraction, calculate $91_{16} - 2C_{16}$.

Minuend	9	$1^{(+16)}$	
Borrow out/in			Subtracted
Subtrahend	$2^{(+1)}$	C	
Difference	6	5	

Therefore $91_{16} - 2C_{16} = 65_{16}$.

Test your knowledge

1 Using the rules for binary addition, evaluate the following:

(a) $1011_2 + 1101_2$ (11000_2)
(b) $1100_2 + 1001_2$ (10101_2)
(c) $1010_2 + 1001_2$ (10011_2)
(d) $1111_2 + 100_2$ (10011_2)
(e) $1110_2 + 110_2$ (10100_2)

2 Using the rules for hex addition, evaluate the following:

(a) $3E_{16} + 2C_{16}$ $(6A_{16})$
(b) $A8_{16} + 9B_{16}$ (143_{16})
(c) $6F_{16} + 29_{16}$ (98_{16})
(d) $BB_{16} + 1A_{16}$ $(D5_{16})$
(e) $47_{16} + 36_{16}$ $(7D_{16})$

3 Using the rules for binary subtraction, evaluate the following:

(a) $1100_2 - 1010_2$ (10_2)
(b) $1001_2 - 101_2$ (100_2)
(c) $1110_2 - 1011_2$ (11_2)
(d) $1001_2 - 11_2$ (110_2)
(e) $1010_2 - 1001_2$ (1_2)

4 Using the rules for hex subtraction, evaluate the following:

(a) $3C_{16} - 2A_{16}$ (12_{16})
(b) $91_{16} - 2C_{16}$ (65_{16})
(c) $60_{16} - 08_{16}$ (58_{16})
(d) $FF_{16} - A9_{16}$ (56_{16})
(e) $A5_{16} - 68_{16}$ $(3D_{16})$

Binary coded decimal (BCD)

Most users of microprocessor-controlled equipment expect inputs and outputs to be in the familiar decimal form. However, for the practical reasons already stated, any such inputs and outputs must be represented in terms of two different states. A special form of binary is therefore used in which each denary digit is represented by its own group of binary digits, and is known as **binary coded decimal** or **BCD**. There are several possibilities for BCD, but that used in a microprocessor is known as **8421 BCD** since these are the weights of each of its binary digits. Each denary digit is expressed as its four-bit pure binary equivalent.

Example 17

Convert the denary number 8576_{10} into its 8241 BCD equivalent.

8	5	7	6
1000	0101	0111	0110

Therefore the BCD equivalent of 8576_{10} is 1000010101110110_2.

Performing arithmetic using BCD is rather awkward, therefore it is usual to convert BCD inputs into pure binary, perform any processing required, and then convert to BCD before displaying the results.

Complements

In any arithmetic the need frequently arises to represent *negative* quantities. In ordinary arithmetic this can be done by using a minus sign, but a microprocessor does not understand this sign. An alternative method must therefore be found to represent negative binary numbers. The usual method of doing this is to make the most significant bit (MSB) of a binary number act as a sign bit with the following meanings:

1   For all **positive numbers**, MSB = 0.
2   For all **negative numbers**, MSB = 1.

This means that for a given number of bits, the magnitude of a signed number is half as great as that for a signed number. For example, an eight-bit binary number may be regarded as:

1   **unsigned**, with values in the range **0 to 255_{10}**, or
2   **signed**, with values in the range **-128_{10} to $+127_{10}$**.

   There are several different methods of representing signed binary numbers using this notation, but the most common method is the use of **complements**. A complement completes a number, that is, a number plus its complement always gives the same result, depending upon the complement used.
   When using a binary system, only two types of complement can be used, the **one's complement** or the **two's complement**. This means that when a binary number and its one's complement are added, the sum of corresponding digits in each number is always 1. Similarly, when adding a number and its two's complement, the sum of corresponding digits in each

number is always 2 (sum of 0, carry out of 1). The two's complement is of particular interest, since when a result is restricted to the lowest 8 or 16 bits, as is the case in a microprocessor, the sum of a number and its two's complement is always zero. Taking the two's complement of a binary number is therefore equivalent to changing the sign of the number.

Complementation is a process that requires relatively simple electronic circuits, and since the only arithmetic that a microprocessor performs is addition, subtraction of a number may be accomplished by adding its complement.

One's complement of a binary number

The one's complement of a binary number is formed by subtracting each digit from 1.

Example 18

Determine the one's complement of 00010111_2 (23_{10}).

$$
\begin{array}{cccccccc}
1 & 1 & 1 & 1 & 1 & 1 & 1 & 1 \\
0 & 0 & 0 & 1 & 0 & 1 & 1 & 1 \\
\hline
1 & 1 & 1 & 0 & 1 & 0 & 0 & 0 \\
\hline
\end{array}
$$

$\left.\phantom{\begin{array}{c}1\\1\end{array}}\right\}$ Subtracted

(one's complement)

Thus the one's complement of $\mathbf{00010111_2}$ is $\mathbf{11101000_2}$.

Example 19

Determine the one's complement of 10101010_2 (170_{10}).

$$
\begin{array}{cccccccc}
1 & 1 & 1 & 1 & 1 & 1 & 1 & 1 \\
1 & 0 & 1 & 0 & 1 & 0 & 1 & 0 \\
\hline
0 & 1 & 0 & 1 & 0 & 1 & 0 & 1 \\
\hline
\end{array}
$$

$\left.\phantom{\begin{array}{c}1\\1\end{array}}\right\}$ Subtracted

(one's complement)

Thus the one's complement of $\mathbf{10101010_2}$ is $\mathbf{01010101_2}$.

It can be seen from these examples that the one's complement of a binary number may be obtained by inversion of all bits, that is, changing all 1's to 0's and all 0's to 1's.

Two's complement of a binary number

The one's complement does not give the true *negative* equivalent of a binary number, and arithmetic using the one's complement will always have an error of 1 unless otherwise corrected. For true representation the **radix complement** must be used, i.e. the two's complement for binary numbers. The two's complement of a binary number may be obtained by adding 1 to the one's complement.

Example 20

Determine the two's complement of 00010111_2 ($+23_{10}$).

1	1	1	1	1	1	1	1	} Subtracted
0	0	0	1	0	1	1	1	

1	1	1	0	1	0	0	0	} Added
							1	

1	1	1	0	1	0	0	1	(two's complement)

Therefore the two's complement of 00010111_2 ($+23_{10}$) is 11101001_2 (-23_{10}).

Note that since the most significant bit (MSB) of 00010111_2 is zero, it may be regarded as a **positive number** and therefore represents $+23_{10}$. Since the most significant bit of its two's complement, 11101001_2, is 1, it may be regarded as a **negative number** and therefore represents -23_{10}.

Example 21

Determine the two's complement of 00111000_2 (56_{10}).

1	1	1	1	1	1	1	1	} Subtracted
0	0	1	1	1	0	0	0	

1	1	0	0	0	1	1	1	} Added
							1	

1	1	0	0	1	0	0	0	(two's complement)

Therefore the two's complement of 00111000_2 ($+56_{10}$) is 11001000_2 (-56_{10}).

A simple method of determining the two's complement of a binary number is to write down all bits from the LSB up to and including the first '1' in true form, and then invert the remaining bits. Using 00111000_2 ($+56_{10}$) as an example:

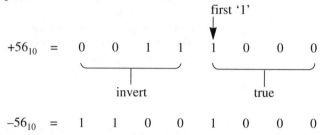

Therefore the two's complement of **00111000_2** ($+56_{10}$) is **11001000_2** (-56_{10}).

Sign extension

When dealing with signed binary, the most significant bit not only indicates the sign of a number, but also the magnitude of any negative part to the number. If the MSB is '0', then the number has no negative part and is a positive number. If it is '1', then the number has a negative part equivalent to the place value of the MSB, and since the MSB has a greater place value than the total of all the remaining bits, the negative part must exceed the positive part and the entire number is negative.

Consider the binary equivalent of -56_{10}:

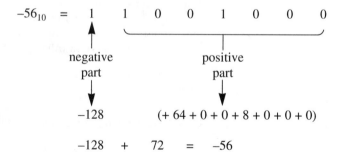

This example shows how the two's complement represents a negative number, and why the MSB must be a '1'. A microprocessor operates with fixed width binary numbers, usually 8, 16 or 32 bits, and therefore numbers that do not require the full number of bits must be **sign extend**ed.

It should be clear that this may be achieved in the case of positive numbers by filling in with leading zeros, a process which in no way alters the value of a number.

Negative numbers, however, must be filled with leading ones. It might be thought that such a process would change the value of a negative number, but this is not so, as can be seen by considering the binary equivalent of -56_{10} with differing numbers of leading ones:

$$-56_{10} = \quad\quad 1 \quad 0 \quad 0 \quad 1 \quad 0 \quad 0 \quad 0$$
$$-64 \;(+\; 0 \;+\; 0 \;+\; 8 \;+\; 0 \;+\; 0 \;+\; 0)$$
$$-64 + 8 = \mathbf{-56}$$

$$-56_{10} = \quad\quad 1 \quad 1 \quad 0 \quad 0 \quad 1 \quad 0 \quad 0 \quad 0$$
$$-128 \;(+64 \;+\; 0 \;+\; 0 \;+\; 8 \;+\; 0 \;+\; 0 \;+\; 0)$$
$$-128 + 72 = \mathbf{-56}$$

$$-56_{10} = \quad 1 \quad 1 \quad 1 \quad 0 \quad 0 \quad 1 \quad 0 \quad 0 \quad 0$$
$$-256 \;(+\; 128 + 64 \;+\; 0 \;+\; 0 \;+\; 8 \;+\; 0 \;+\; 0 \;+\; 0)$$
$$-256 + 200 = \mathbf{-56}$$

Hex complements

It is generally a simple matter to process hex numbers by converting to binary, complementing and converting back, but it is possible to complement hex numbers directly. As might be expected, the **15's complement** or the **16's complement** is used for hex numbers, and the conversion processes are similar to those used for binary numbers.

15's complement of a hex number

The 15's complement of a hex number is obtained by subtracting each digit from F_{16} (15_{10}).

Example 22

Determine the 15's complement of 17_{16} (23_{10}).

$$\left.\begin{array}{cc} F & F \\ 1 & 7 \end{array}\right\} \text{Subtract}$$

$$\begin{array}{cc} \hline E & 8 \end{array} \quad \text{(15's complement)}$$

Thus the 15's complement of $\mathbf{17_{16}}$ is $\mathbf{E8_{16}}$.

Example 23

Determine the 15's complement of 38_{16} (56_{10}).

$$\left.\begin{array}{cc} F & F \\ 3 & 8 \end{array}\right\} \text{Subtract}$$

$$\begin{array}{cc} \hline C & 7 \end{array} \quad \text{(15's complement)}$$

Thus the 15's complement of $\mathbf{38_{16}}$ is $\mathbf{C7_{16}}$.

Since these examples are identical to the binary examples previously considered, the results may be readily compared. A similar problem occurs as with binary one's complement, and for true representation of negative numbers the **radix complement**, or **16's complement**, must be used. The 16's complement may be obtained by adding one to the 15's complement, or directly by subtraction of the LSD from 10_{16} (16_{10}) and the remaining digits from F_{16} (15_{10}).

Example 24

Determine the 16's complement of 17_{16} (23_{10}).

$$\left.\begin{array}{cc} F & F \\ 1 & 7 \end{array}\right\} \text{Subtract} \qquad \left.\begin{array}{cc} F & 10 \\ 1 & 7 \end{array}\right\} \begin{array}{l} (10_{16} = 16_{10}) \\ \text{Subtract} \end{array}$$

$$(15\text{'s}) \quad \left.\begin{array}{cc} E & 8 \\ & 1 \end{array}\right\} \text{Add} \qquad (16\text{'s}) \quad E \quad 9$$

$$(16\text{'s}) \quad E \quad 9$$

Thus the 16's complement of $\mathbf{17_{16}}$ is $\mathbf{E9_{16}}$.

Example 25

Determine the 16's complement of 38_{16} (56_{10}).

$$\left.\begin{array}{cc} F & F \\ 3 & 8 \end{array}\right\} \text{Subtract} \qquad \left.\begin{array}{cc} F & 10 \\ 3 & 8 \end{array}\right\} \begin{array}{l} (10_{16} = 16_{10}) \\ \text{Subtract} \end{array}$$

$$(15\text{'s}) \quad \left.\begin{array}{cc} C & 7 \\ & 1 \end{array}\right\} \text{Add} \qquad (16\text{'s}) \quad C \quad 8$$

$$(16\text{'s}) \quad C \quad 8$$

Thus the 16's complement of 38_{16} is $C8_{16}$.

Example 26

Convert -50_{10} to hexadecimal.

This calculation may best be performed by first converting $+50_{10}$ to hex, then determining the 16's complement of the result.

Convert $+50_{10}$ to hex:

30/16	=	3	remainder 2
3/16	=	0	remainder 3

Therefore $+50_{10} = 32_{16}$.

Determine the 16's complement:

$$\left. \begin{array}{cc} F & F \\ 3 & 2 \end{array} \right\} \text{Subtract}$$

$$\left. \begin{array}{cc} C & D \\ & 1 \end{array} \right\} \text{Add}$$

$$\begin{array}{cc} C & E \end{array}$$

Therefore $-50_{10} = CE_{16}$.

Subtraction using complements

A microprocessor can only *add* numbers. Fortunately subtraction can be performed by the addition of a negative number; for example:

$12 - 7$ may be written as $12 + (-7)$

It should now be clear that a microprocessor may perform subtraction by adding the complement of the subtrahend to the minuend. This may be achieved by adding the one's complement provided an **end around carry** is used, or by adding the two's complement and ignoring any **spill over** outside of the maximum number of bits used.

Example 27

Perform the subtraction $01100_2(+12_{10}) - 00101_2(+5_{10})$ by using the one's complement.

Minuend $(+12_{10})$ = 01100   Subtrahend $(+5_{10})$ = 00101

                      + 11010 ◄— One's complement = 11010

Sum	$\boxed{1}$	00110
End around carry		➤ 1
Sum		00111

$\left. \right\}$ Add

Thus $01100_2 - 00101_2 = 00111_2(7_{10})$.

Example 28

Perform the subtraction $01100_2(+12_{10}) - 00101_2(+5_{10})$ by using the two's complement.

Minuend $(+12_{10})$	=	01100	Subtrahend $(+5_{10})$	=	00101
Spill over (ignore) $\boxed{1}$	$+$	11011 $\longleftarrow$ Two's complement		=	11011
Sum		00111			

Thus $\mathbf{01100_2 - 00101_2 = 00111_2}(7_{10})$.

   Note that it is essential to sign extend to make the width of each number identical.

Example 29

Perform the subtraction $93_{16}\,(147_{10}) - 2C_{16}\,(44_{10})$ using the 15's complement method.

$$
\begin{array}{lll}
 & & \text{F} \quad \text{F} \\
 & \text{Subtrahend}(44_{10}) \;=\; & 2 \quad \text{C}
\end{array}\Bigg\} \text{Subtract}
$$

Minuend $(+147_{10})$	=	9	3	
	+	D	3 $\longleftarrow$ 15's complement	= D 3
Sum	$\boxed{1}$	6	6	
End around carry		$\longrightarrow$ 1		
Sum		6	7	

Thus $\mathbf{93_{16} - 2C_{16} = 67_{16}}\,(\mathbf{103_{10}})$.

Example 30

Perform the subtraction $A1_{16}(161_{10}) - 5B_{16}(91_{10})$ using the 16's complement method.

$$
\begin{array}{rcl}
 & & \left. \begin{array}{cc} F & F \\ 5 & B \end{array} \right\} \text{Subtract} \\
\text{Subtrahend } (+91_{10}) = & & \\
 & & \\
\text{15's complement} \quad = & & \left. \begin{array}{cc} A & 4 \\ & +1 \end{array} \right\} \text{Add} \\
\end{array}
$$

Minuend $(+161_{10})$ =   A   1

Spill over (ignore) $\boxed{1}$ A   5 ◄— 16's complement   A   5

Sum   4   6

Thus $A1_{16} - 5B_{16} = 46_{16}$ (70_{10}).

Test your knowledge

1   Determine the one's complement of the following binary numbers:

    (a) 11010111_2 (00101000_2)
    (b) 01101110_2 (10010001_2)
    (c) 11100011_2 (00011100_2)
    (d) 10110110_2 (01001001_2)
    (e) 10001000_2 (01110111_2)

2   Determine the two's complement of the following binary numbers:

    (a) 10011100_2 (01100100_2)
    (b) 11110000_2 (00010000_2)
    (c) 10101010_2 (01010110_2)
    (d) 11011100_2 (00100100_2)
    (e) 10100111_2 (01011001_2)

3   Determine the 15's complement of the following hex numbers:

    (a) $2D_{16}$ $(D2_{16})$
    (b) FF_{16} (00_{16})
    (c) 10_{16} (EF_{16})
    (d) CB_{16} (34_{16})
    (e) $7E_{16}$ (81_{16})

4 Determine the 16's complement of the following hex numbers:

(a) 02_{16} (FE_{16})
(b) $2F_{16}$ $(D1_{16})$
(c) $8B_{16}$ (75_{16})
(d) AF_{16} (51_{16})
(e) $F8_{16}$ (08_{16})

5 Express the following denary numbers as signed binary:

(a) -25_{10} (100111_2)
(b) $+3_{10}$ (011_2)
(c) -100_{10} (10011100_2)
(d) $+31_{10}$ (011111_2)
(e) -89_{10} (10100111_2)

6 Express the following denary numbers as signed hex:

(a) $+8_{10}$ (51_{16})
(b) -36_{10} (DC_{16})
(c) $+125_{10}$ $(7D_{16})$
(d) -62_{10} $(C2_{16})$
(e) -120_{10} (88_{16})

7 Using complements, evaluate the following:

(a) $011011_2 - 01110_2$ (001101_2)
(b) $011110_2 - 1101_2$ (0100001_2)
(c) $6A_{16} - 2F_{16}$ $(3B_{16})$
(d) $1E_{16} - FE_{16}$ (20_{16})
(e) $BB_{16} - 8F_{16}$ $(2C_{16})$

2 Microprocessor-based systems

Summary

This element introduces the concept of using microcomputers as system controllers, and describes a microcomputer in terms of its constituent parts at a block diagram level. The functions and operations of the individual elements of a microprocessor are first described in isolation, then how they work together by studying the fetch–execute cycle. Different types of machine cycle are identified and illustrated using basic timing diagrams. Three different types of microprocessor, the Z80, 8086 and 8031, are used as examples in the text.

The microcomputer   Systems

A system may be defined as an orderly arrangement of physical or abstract objects. Systems have **inputs** and **outputs** arranged as shown in Figure 2.1.

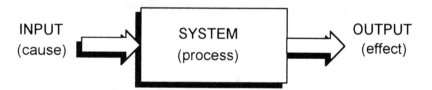

Figure 2.1

The input signal may cause the system output to change or may cause the operation of the system to change. Therefore the input signal is the **cause** of the change and the resulting action is called the **effect**. The response of the system to an input signal is called the **process**.

Integrated circuits

The processing part of a system may be electronic, ranging from very simple to highly complex circuits. Simple circuits may be constructed using individual (or discrete) components such as transistors, resistors, capacitors etc., but in more complex circuits **integrated circuits** (ICs) are used.

23

Integrated circuits are complete circuits that are manufactured by a common process rather than being assembled from individual components. Many different types of integrated circuit are manufactured to perform one type of process only. This means that when a different process is required, a different type of IC must be used, and other changes to the circuit may be necessary. A **microprocessor** is a type of IC that avoids this problem because it is not dedicated to one particular process, but behaves as instructed by a sequence of binary numbers applied to its inputs. This sequence of binary instructions is called a **program**, and it is this that determines the characteristics of a microprocessor-based system.

Microcomputer block diagram

A microprocessor program must be stored in memory so that each instruction is readily available as and when required. When a microprocessor and its program memory are connected together, a microcomputer is formed. Input and output facilities are also required in a practical **microcomputer**, as shown in Figure 2.2.

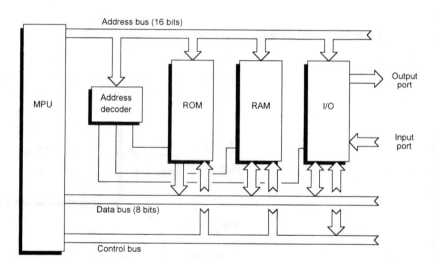

Figure 2.2

The functions of each part of the microcomputer in Figure 2.2 are as follows.

The microprocessing unit or MPU

The **microprocessing unit** (MPU) or **central processing unit** (CPU) is responsible for executing a wide variety of data transfer, arithmetic and logic instructions, under program control, and for generating all control signals required to achieve this. It therefore acts as the *processing* or *controlling* element of a microprocessor-based system.

Read-only memory (ROM)

This is a memory (or store), the contents of which are fixed during manufacture. The contents of a ROM are retained even after switching off its power supply. As its name implies, the contents of this memory may be read by an MPU, but cannot be altered. It is therefore used to store data and MPU instructions which, for a given system, are fixed and never need changing. Instructions stored in ROM are called **firmware**.

Random access memory (RAM)

This is a memory which is more correctly described as a read/write memory. Information may be read from and written (i.e. stored) into this type of memory. It is therefore used to store information of a changeable nature, e.g. results of calculations, user's programs. Information stored in most types of RAM is lost after switching off its power supply. Instructions stored in RAM are known as **software**.

Input/output (I/O)

Since a microcomputer is the processing part of a system, a facility to enable communication with input and output devices (peripherals) is essential. This is provided by the input/output (I/O) section of the microcomputer which enables input and output signals to be transferred between peripheral devices and the data bus.

Address bus

Each storage location within a memory device must have a unique identification so that the system is able to select a particular instruction or data item for processing. This identification is called an **address** and takes the form of a binary number (usually 16 or more bits). The address bus consists of a parallel group of conductors which convey addresses generated by the MPU to other components in the microcomputer. It is therefore a **unidirectional bus**.

Address decoder

A microcomputer may be constructed using several different types of memory or I/O device, each of which must be selected when needed, so that data may be directed to or from the correct device. Not all of the available bits on the address bus are needed to identify a memory or I/O location, therefore it is usual for the higher order bits of each address to contain information that identifies the device to be selected. An address decoder is used to generate appropriate device select signals from each address.

Data bus

During execution of a microcomputer program, information in the form of parallel groups of data bits must be transferred between an MPU and its memory and input/output devices. The **data bus** consists of a parallel group of conductors that enable such transfers to take place. Since transfers may take place both into and out of the MPU, it must be a **bidirectional bus**.

Control bus

A microcomputer requires a number of signals for its correct operation whose purpose is to control operations within the system. For example, control signals are required to reset a microcomputer, or switch a RAM between read and write modes. These individual signals are, for convenience, grouped together and conveyed by means of a control bus.

Test your knowledge 2.1

1   An orderly collection of physical or abstract objects is known as:
A   a microcomputer
B   a system
C   an integrated circuit
D   a process

2   A programmable device that may be used to control a system is:
A   an integrated circuit
B   a discrete circuit
C   a dedicated circuit
D   a microprocessor

3   The permanent storage of fixed data in a microcomputer is
    achieved by the use of a:
    A   RAM
    B   MPU
    C   ROM
    D   CPU

4   The temporary storage of variable data in a microcomputer
    is achieved by the use of a:
    A   RAM
    B   MPU
    C   ROM
    D   CPU

5   An MPU processes information supplied to it via a:
    A   bidirectional data bus
    B   bidirectional address bus
    C   unidirectional data bus
    D   unidirectional address bus

6   An MPU generates a binary number to identify one
    particular memory location and puts this out onto the:
    A   data bus
    B   control bus
    C   address bus
    D   peripheral bus

7   The I/O section of a microcomputer allows an input
    peripheral to communicate with the:
    A   MPU via the address bus
    B   ROM via the data bus
    C   ROM via the address bus
    D   MPU via the data bus

8   A microcomputer address decoder is used to decode:
    A   low order address lines to select a memory device
    B   high order address lines to select a memory location
    C   high order address lines to select a memory device
    D   low order address lines to select a memory location

Microprocessor block diagram

Most types of MPU operate in a similar manner, but their internal structure
(or *architecture*) may vary considerably. Certain features are common to all
types, and these are shown in Figure 2.3.

Registers

It can be seen from Figure 2.3 that an MPU contains a number of different
registers that provide storage for multi-bit data during processing. A single

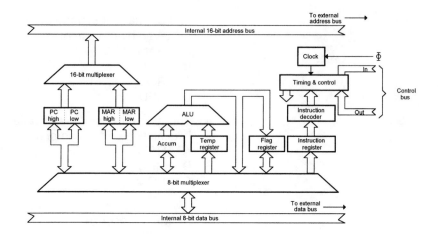

Figure 2.3

bit (1 or 0) may be stored in a bistable (or flip-flop) circuit, and a register consists of a number of these circuits operating with a common clocking signal (see Figure 2.4). A register typically has 8, 16 or 32 bistable circuits, depending upon the purpose of the register and the type of MPU. A *counter* is a register, organized such that each clock pulse causes the contents of the register to be incremented by one.

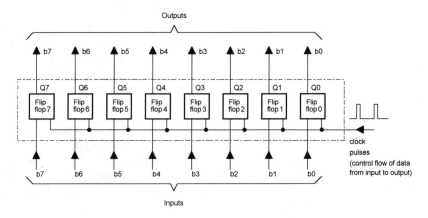

Figure 2.4

Program counter (PC)

Instructions must be executed in the order in which they appear in a program (except when changes in this sequence are ordered by the program itself). The **program counter**, or **PC**, is a 16-bit register that holds the address of the next byte of the stored program sequence. In order to fetch the next program byte from memory, the PC contents must first be copied onto the address bus. This action automatically causes the address in the PC to be incremented by one. The PC is connected to the address bus via a **multiplexer/latch**, the outputs of which hold the current address steady on the bus while the next program address is prepared by incrementing the PC. This arrangement is shown in Figure 2.5.

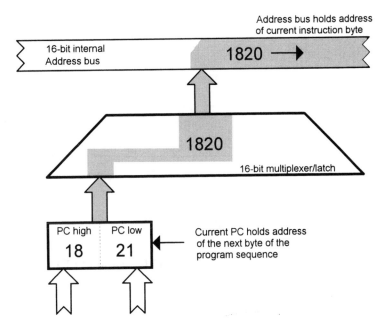

Figure 2.5

Instruction register (IR)

The first byte of each instruction informs an MPU what *type of operation* to carry out (e.g. add, subtract etc.). This information appears on the data bus for only a very short period of time, and must therefore be captured and stored to allow time for its subsequent decoding. The **instruction register** is responsible for this capture and store process. Each instruction has a different binary pattern containing information that specifies the type of operation and defines the registers to use. This binary pattern must be analysed (or decoded) to enable an MPU to select the correct registers and perform the correct operation.

The instruction decoder analyses the contents of the instruction register and passes its decoded outputs to the timing and control section for implementation of the instruction. This process is shown in Figure 2.6 which uses the Z80 instruction ADD E as an example (*add contents of register E to the accumulator*).

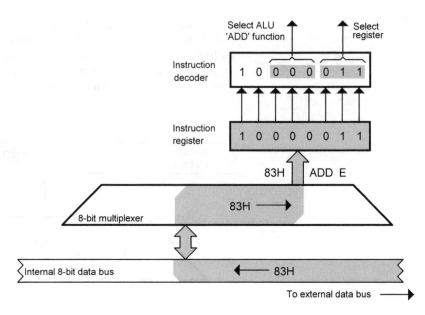

Figure 2.6

Arithmetic and logic unit (ALU)

Microprocessors are typically classified as 4-, 8-, 16- or 32-bit processors, according to the number of bits that can be simultaneously processed. For example, an 8-bit MPU processes eight bits simultaneously. The number of bits that can be processed simultaneously is determined by the number of **binary adder** circuits in the arithmetic and logic unit (ALU) of an MPU. The ALU is the heart of the processing system, and it performs all **arithmetic** and **logic** processes within an MPU. An 8-bit MPU performs operations between two 8-bit numbers and delivers a corresponding result. Arithmetic and logic processes almost always involve processing two numbers (or *operands*), e.g. addition or subtraction of two numbers. Therefore the ALU in an 8-bit MPU has two 8-bit input ports, one for each operand. Central to every ALU is a **parallel binary adder chain**, organized as shown in Figure 2.7.

Accumulator (A)

The two ALU operands must each be stored in registers while the ALU performs its operation on them. One of these registers is unique in that the result of the operation is stored back into it to replace its original contents. This register is called the **accumulator** and is used directly by many instructions. The second register is used only for short-term storage of one of the operands during processing and is called a **temporary register**. The

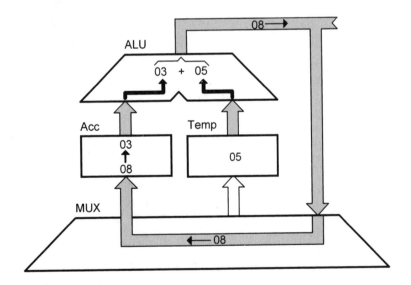

Figure 2.7

behaviour of the accumulator and temporary register during a typical instruction is shown in Figure 2.8.

Figure 2.8

Flag register (F)

It is important to get information about the *result* of an arithmetic or logic operation, in addition to its actual value. This is particularly necessary when making decisions regarding the direction of program flow, or for general manipulation and testing of numbers. An MPU contains a number of individual **bistables** (or **flip-flops**) called **flags** that indicate when specific conditions occur. For convenience these flags are grouped together and are collectively known as a **flag register** or **status register**. Most MPUs provide the following four flags plus other more specialized flags:

1 **Zero flag**   (Z or ZF)
2 **Carry flag**   (C, CY or CF)
3 **Sign flag**   (S or SF)
4 **Overflow flag**   (V, OV or OF)

The operation of these flags is shown in Table 2.1.

Table 2.1

Flag	Clear	Set
Z	result not zero	result is zero
C	unsigned result 0 <= RESULT <= 255	unsigned result 0 > RESULT > 255
S	result POSITIVE (bit 7 = 0)	result NEGATIVE (bit 7 = 1)
V	sign correct −128 <= RESULT <= +127	sign incorrect −128 > RESULT > +127

Example

Determine the state of the Z, C, S and V flags of an 8-bit microprocessor when it carries out the following operations:

(a) 02 + 05
(b) F9 + 08
(c) 30 + 70
(d) E0 + 20

(a) 02 + 05 = 07
    The result is **not zero**, therefore **Z = 0**
    The unsigned result is **less than 256**, therefore **C = 0**
    The result is **positive**, therefore **S = 0**
    The signed result is **within the range −128 to +127**, therefore **V = 0**

(b) F9 + 08 = (1)01
    The result is **not zero**, therefore **Z = 0**
    The unsigned result is **greater than 256**, therefore **C = 1**
    The result is **positive**, therefore **S = 0**
    The signed result is **within the range −128 to +127**, therefore **V = 0**

(c) 30 + 70 = A0
The result is **not zero**, therefore **Z = 0**
The unsigned result is **less than 256**, therefore **C = 0**
The result is **negative**, therefore **S = 1**
The signed result is **outside the range –128 to +127**, therefore **V = 1**

(d) E0 + 20 = (1)00
The result is **zero**, therefore **Z = 1**
The unsigned result is **greater than 256**, therefore **C = 1**
The result is **positive**, therefore **S = 0**
The signed result is **inside the range –128 to +127**, therefore **V = 0**

Memory address register (MAR)

A program counter generates a sequence of addresses which allow program instructions to be fetched from memory and executed in the correct order. During the execution of a program, however, it may be necessary to access data at addresses other than those in the program sequence. A program counter cannot be used to generate addresses for this purpose since this would result in loss of the address of the next program byte. The generation of such out of sequence addresses is achieved with the aid of a 16-bit register called a **memory address register**. In systems using an 8-bit data bus, addresses require two fetches from memory, one for the lower eight bits and another for the upper eight bits. Therefore a memory address register is often loaded as two separate 8-bit registers which are combined into a single 16-bit register when delivering addresses. This register automatically has the required addresses stored in it as part of the execution sequence of certain instructions (see Figure 2.9).

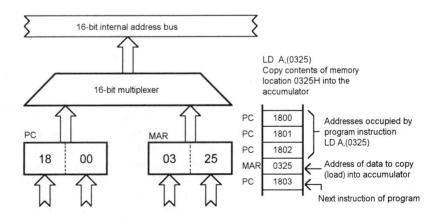

Figure 2.9

General purpose registers

In addition to the specialized memory address register already described, an MPU will often contain a number of general purpose registers. These may be used for holding either data or addresses, and may be treated as individual 8-bit registers, or may be combined as a 16- or 32-bit register for data or addresses (see Figure 2.10).

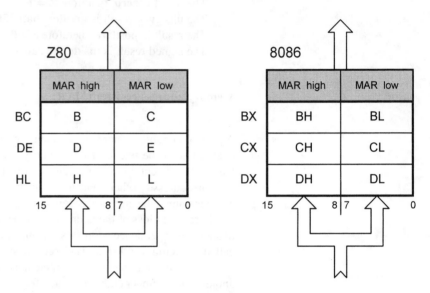

Figure 2.10

The availability of general purpose registers will vary from one type of microprocessor to another. Figure 2.10 shows arrangements for the Z80 and 8086 MPUs.

Multiplexer (MUX)

A multiplexer is a switching circuit that allows several different signal sources to *share* a common data path. Different signals obviously cannot be simultaneously connected to a data path, otherwise they would interfere with one another. Therefore each signal is allocated a unique time slot during which the data path is available for its exclusive use. In an MPU, internal data and address buses are shared by various registers, and during each operation these are switched onto the buses, as and when necessary, by means of multiplexer circuits (see Figure 2.11).

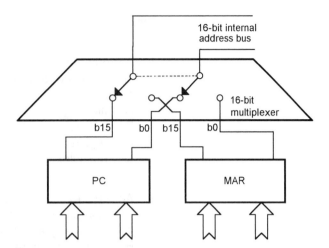

Figure 2.11

Control unit

In order for an MPU to operate in a logical manner, as instructed, it is necessary to coordinate all internal circuits and operations, e.g. select correct registers, transfer data at correct time, etc. It is also necessary for an MPU to generate all external control signals required by the memory and I/O circuits, and to respond correctly to various input control signals. The control unit ensures that this coordination takes place, and it consists of a ROM sequencer circuit that generates a very large number of individual internal and external control signals, timed by a common system clock. Each MPU instruction has its own particular control signal sequence. This arrangement is shown in Figure 2.12.

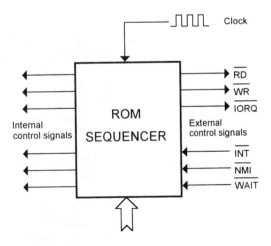

Figure 2.12

Test your knowledge 2.2

1   A group of bistable circuits that share a common clock, and are used for storing groups of data bits, are known as a:
A   flip-flop
B   counter
C   register
D   multiplexer

2   An MPU executes instructions in the correct sequence, because program addresses are supplied by:
A   a memory address register
B   an instruction decoder
C   an instruction register
D   a program counter

3   The results of arithmetic calculations in a microprocessor are stored in the:
A   accumulator
B   arithmetic and logic unit
C   temporary register
D   flag register

4   The first byte of every instruction is stored in the:
A   accumulator
B   instruction decoder
C   arithmetic and logic unit
D   instruction register

5   A single bit store that is used to indicate when a particular condition occurs within a microprocessor is known as:
A   a register
B   a flag
C   an accumulator
D   a decoder

6   If the result of an arithmetic calculation is an 8-bit number in the range 01 to 7F, then the states of the Z and S flags are:
A   Z=0, S=0
B   Z=0, S=1
C   Z=1, S=0
D   Z=1, S=1

7   A register that provides addresses outside of the normal
    program sequence for the purpose of accessing data is
    known as:
    A   an instruction register
    B   an accumulator
    C   a temporary register
    D   a memory address register

8   In order for an MPU to operate in a coordinated manner,
    internal and external timing and control signals are
    generated by the:
    A   instruction register
    B   multiplexer
    C   ROM sequencer circuit
    D   instruction decoder

Activity

1   Obtain instructions in the use of a microprocessor
    debugging program so that you are able to carry out
    the following operations:
    (a)   enter programs using machine code or
          assembly language,
    (b)   inspect/alter the program data in memory,
    (c)   inspect/alter the microprocessor registers,
    (d)   trace program execution one instruction at a
          time.

2   Determine theoretical results and effect on the Z,C,S
    and V flags of the operations shown in Table 2.2 and
    record your results.

Table 2.2

Operand 1	Operation	Operand 2	Acc	Z	C	S	V
05	ADD	08					
60	ADD	30					
07	ADD	F9					
82	ADD	80					
0C	SUB	03					
60	SUB	80					
32	SUB	CE					
40	SUB	60					

3 Write simple programs to test the results of each calculation, e.g.

```
Z80    ld     a,05       ld     a,0c
       add    a,08       sub    03

8086   mov    al,05      mov    al,0c
       add    al,08      sub    al,03

8031   mov    a,#05      clr    c      (CY and OV only)
       add    a,#08      mov    a,#0c
                         sub    a,#08
```

(Suitable program endings must be added after each program, but depend very much upon the equipment used.)

4 Enter these programs into your microprocessor system, step through them, inspect the registers and check that the predicted results are obtained, and if not, recheck calculations.

Practical microprocessors

A very wide range of different microprocessors is available, and selecting a microprocessor for a particular application will depend upon the requirements of the system into which it is incorporated. Factors such as **cost, speed** of operation, **size** and **power consumption** will all have to be considered. For example, a high performance desktop computer requires a very fast and relatively expensive microprocessor, but for a simple control application, low cost rather than high speed may be the most important factor. Three contrasting types of microprocessor are shown in Figures 2.13, 2.14 and 2.15.

Note that the **8086** is used to represent the **Intel iAPX86** range of MPUs, since the later (and more advanced) versions are compatible with this device. Similarly, the **8031** is representative of one of the **Intel MCS-51** range of single chip microcontrollers.

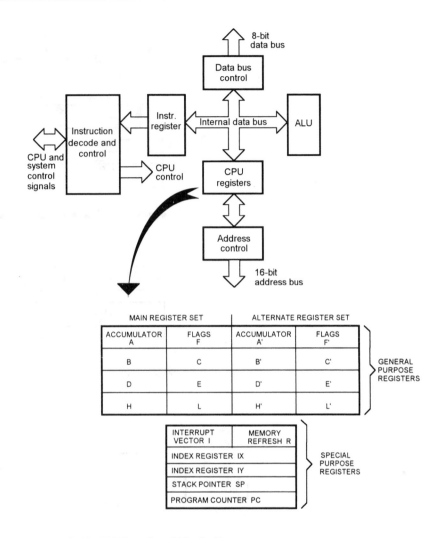

Figure 2.13 Z80 functional block diagram

Activity

Write a report in which two microprocessors, one from the 8031 family and one from the 8086 family, are compared, explaining why the 8031 is more suitable for simple control applications whereas the 8086 is more suitable for use in desktop microcomputers.

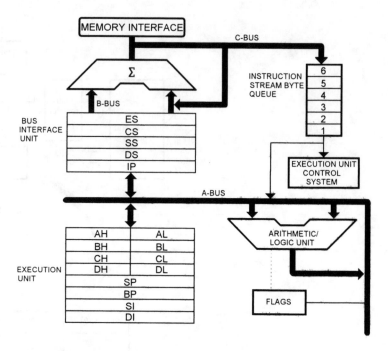

Figure 2.14 8086 functional block diagram

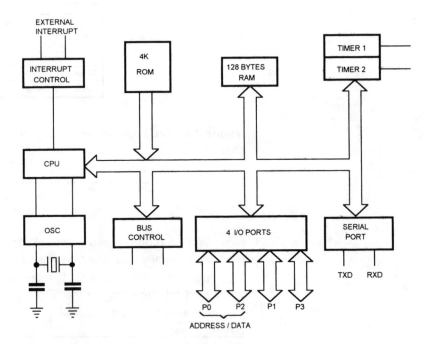

Figure 2.15 Block diagram MCS-51 series

The fetch–execute cycle

A microprocessor program consists of a *sequence of instructions* which, when carried out in the correct order, enable a microcomputer to perform its required functions. Each instruction consists of one or more bytes of machine code with the following functions:

Operator: the first one or two bytes of an instruction that define the exact **operation** to be performed by a microprocessor, e.g. ADD, SUBTRACT, INCREMENT.

Operand: the following one or more bytes that specify the **data** (or location of data) upon which the instruction operates.

A typical instruction for the Z80 8-bit MPU copies the contents of a memory location (e.g. 1825H) into the accumulator, and may be written as shown in Figure 2.16.

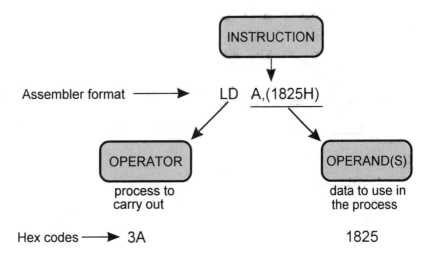

Figure 2.16

A microprocessor carries out each instruction by performing a sequence of operations, known as the **fetch–execute cycle**, that consists of the following steps:

1   **Fetch** the operator code (opcode) from memory and store it in the microprocessor's **instruction register** (IR).
2   **Decode** the opcode within the microprocessor's **instruction decoder** unit (IDU) to determine the nature of the operation being specified by the instruction.
3   **Fetch** further data if necessary.
4   **Execute** the instruction.

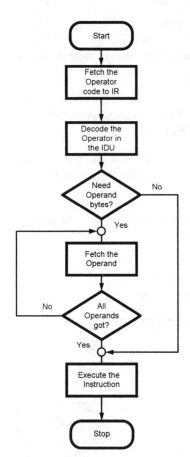

Figure 2.17

The fetch–execute cycle may be represented as a flowchart as shown in Figure 2.17.

Typical fetch–execute cycle

A Z80 instruction that may be used to study a typical fetch–execute cycle is:

LD A,(nn)

which means '*copy the contents of memory address "nn" into the accumulator*', i.e. load accumulator from memory. If, in this example, address '*nn*' is 1825H, then the instruction consists of the following machine code sequence:

3A 25 18

If this instruction is stored in memory, starting at address 1800H, then it is represented in the following manner:

1800   3A 25 18   LD   A,(1825H)

and is stored in memory as shown in Table 2.3.

Table 2.3

Address	Data
1800	3A
1801	25
1802	18
1803	★★
…	…

The fetch–execute cycle then consists of the following four operations:

1 **Fetch the opcode** (3A) from memory via the data bus, and transfer it into the instruction register (see Figure 2.18(a)).
2 **Fetch the low byte** (25) of the external address from memory, and transfer it to the lower eight bits of the MAR (see Figure 2.18(b)).
3 **Fetch the high byte** (18) of the external address from memory, and transfer it to the upper eight bits of the MAR (see Figure 2.18(c)).
4 Put the address in the MAR (1825) out onto the address bus, and transfer the data (55) from this address, via the data bus, into the accumulator (see Figure 2.18(d)).

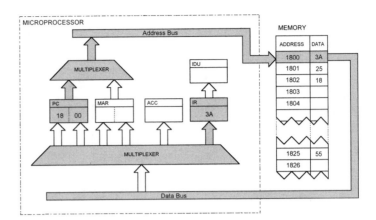

Figure 2.18(a)

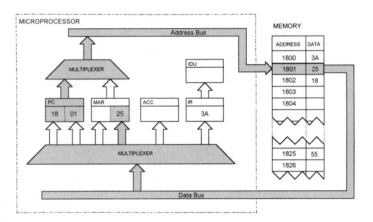

Figure 2.18(b)

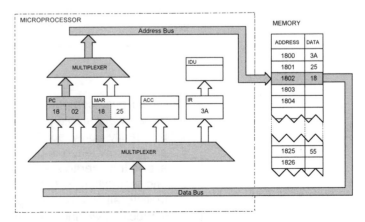

Figure 2.18(c)

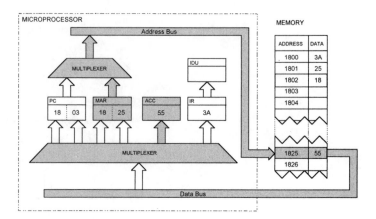

Figure 2.18(d)

This sequence may be summarized as shown in Table 2.4.

Table 2.4

Step	Action		
1	PC	→	Address Bus
2	PC+1	→	PC
3	Data Bus	→	IR
4	PC	→	Address Bus
5	PC+1	→	PC
6	Data Bus	→	MAR low
7	PC	→	Address Bus
8	PC+1	→	PC
9	Data Bus	→	MAR high
10	MAR	→	Address Bus
11	Data Bus	→	Accumulator

Activity

If your microcomputer system has a facility to *single cycle* or *hardware step* a program then the fetch – execute may be readily analysed. Do not confuse this with the *single step* facility often found in debugging programs, which allows the user to step through a program from the start of one instruction to the start of the next. A hardware step facility allows a microprocessor to step through an instruction, waiting at each point in the fetch–execute cycle until hardware stepped, normally by the user pressing a certain key. The microprocessor waits at each point in the instruction where bus conditions are stable, therefore it is possible to monitor the state of the buses during each part of the fetch–execute cycle, using a logic probe or an LED indicator circuit (already built into some types of microcomputer trainer).

The following example shows the steps necessary to check the fetch–execute cycle of a Z80 system. This process may be adapted for other microprocessors.

1 Check that the **hardware step** switch is **off**, then starting at address 1800H, enter the following program:
LD   A,(1825H)
LD   (1826H),A
JP   1800H
2 Construct a table similar to Table 2.5 and enter the corresponding hex codes for each instruction.

Table 2.5

Instruction		Hex codes		
LD A,(1825H)				
LD (1826H),A				
JMP 1800H				

3 Use the **modify memory** command to store hex number 55 at address 1825H.
4 Run the program, then switch on the **hardware step**.
5 Hardware step repeatedly until the address bus monitor LEDs (A0–A15) show the binary equivalent of 1800H, and the data bus monitor LEDs (D0–D7) show the binary equivalent of 3AH.
6 Construct a table similar to Table 2.6. Using the 'hardware step' key, hardware step the program and enter the results in your table.
7 Identify the **fetch** and **execute** cycles and mark this onto the table.

Table 2.6

Instruction	Address	Data	RD	WR	Operation

Timing diagrams

A timing diagram shows, *in correct time sequence*, the states of all relevant bus and control lines during a specified operation. The condition of data on any bus or control line may at any instant be either valid, invalid or uncertain. Examples of each of these conditions are shown in Figure 2.19.

Signals involved in memory read and memory write operations

The signals required to access a memory system for the purpose of read and write operations are similar to the following Z80 signals:

A0–A15	used to select the memory location to be accessed
D0–D7	data for transfer into memory (write) or out of memory (read)
$\overline{\text{MREQ}}$	enables memory system rather than the I/O system for data transfer
$\overline{\text{RD}}$	puts memory into the read mode
$\overline{\text{WR}}$	puts memory into the write mode

All of these signals (other than D0 to D7 during memory read) are generated directly, or indirectly, by the MPU. During a memory read, D0 to D7 are supplied by the memory system itself. The 8086 and 8031 series of microprocessors use *multiplexing* techniques in which data and addresses share the same pins of the MPU. These MPUs require another signal, called

ALE (address latch enable), to separate out the addresses from the data. Typical arrangements for the Z80, 8086 and 8031 are shown in Figures 2.20, 2.21 and 2.22.

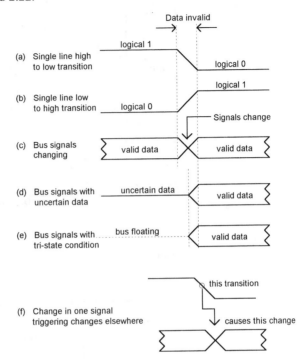

Figure 2.19

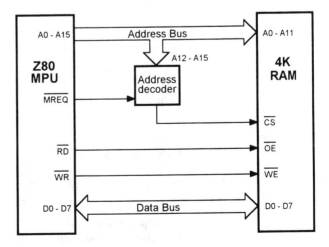

Figure 2.20

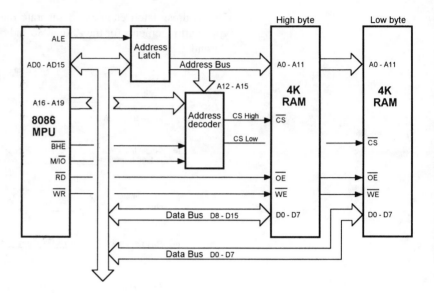

Figure 2.21

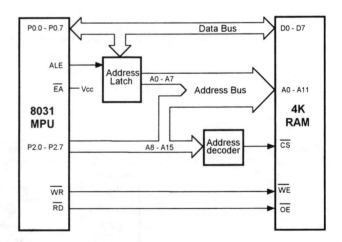

Figure 2.22

MPU basic operations

The different types of basic operations performed by a microprocessor are limited, the most common being:

1 **Opcode fetch** cycle
2 **Memory read** cycle
3 **Memory write** cycle

4   **Input read** cycle
5   **Output write** cycle

All instructions consist of a sequence of these basic operations, each operation taking at least 3–4 cycles of the system clock to complete (12 for the 8031).

All instructions start with an **opcode fetch** (OCF) which is in most cases similar to a memory read cycle. **Memory read** (MR) and **memory write** (MW) cycles may then follow, depending upon the exact nature of the instruction.

Each clock period is often called a **'T' (time) state**, and each basic operation is called a **bus cycle** (or machine cycle). The relationships between T states and bus cycles are shown in Figure 2.23.

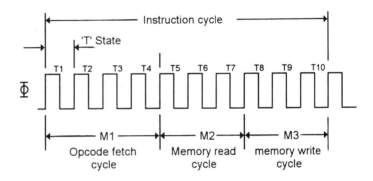

Figure 2.23

Read bus cycle (simplified)

A basic read cycle is shown in Figure 2.24. To read data from a memory location into the MPU, the following sequence of operations is carried out:

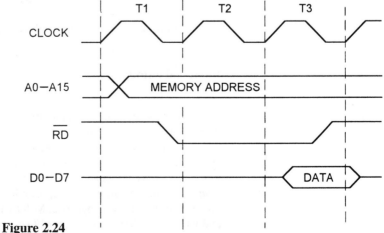

Figure 2.24

During T1: put an address out onto the address bus.
During T2: generate active $\overline{RD}$ (read memory) and $\overline{MREQ}$ signals.
During T3: wait for memory to put addressed data onto data bus.
End of T3: data on data bus clocked into MPU.

Write bus cycle (simplified)

A basic write cycle is shown in Figure 2.25. To write data from MPU into a memory location, the following sequence of operations is carried out:

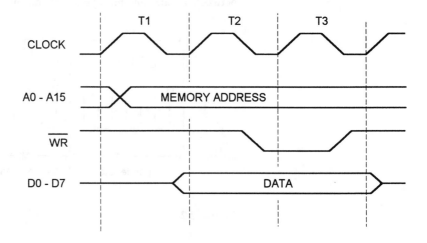

Figure 2.25

During T1: put an address out onto the address bus.
During T3: generate active $\overline{WR}$ (read memory) and $\overline{MREQ}$ signals; MPU outputs data to be written into memory onto the data bus.
End of T3: data from data bus is clocked into memory.

Input and output cycles

I/O read and write operations are often distinguished from memory read and write operations by means of special control signals and by the use of IN and OUT instructions. For example, the Z80 MPU uses $\overline{MREQ}$ and $\overline{IORQ}$ signals to identify memory and I/O operations. During memory read and write operations $\overline{MREQ}$ is held low and $\overline{IORQ}$ is held high, while during I/O read and write operations the opposite logic levels apply. The 8086 has only one control line for this purpose, $M/\overline{IO}$ which is held high for memory read and write operations, and low for I/O operations. When memory and I/O are separated in this way, an MPU is said to use **isolated I/O**. The 8031 microcontroller makes no such distinction between memory and I/O and the peripheral devices are said to be **memory mapped**. Therefore no special signals are used in this device to differentiate between memory and I/O.

Practical timing diagrams

Timing diagrams for the Z80 opcode fetch, memory read, memory write, input and output cycles are shown in Figures 2.26(a)–(d). Note the use of

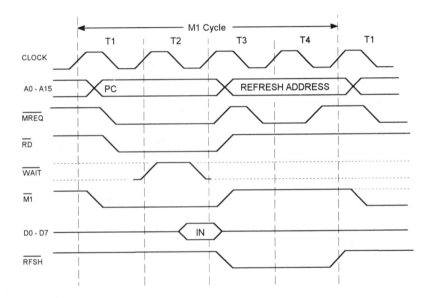

Figure 2.26(a) Z80 opcode fetch cycle

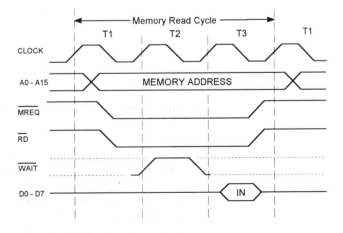

Figure 2.26(b) Z80 memory read cycle

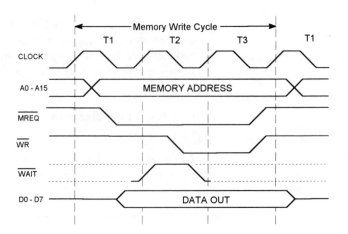

Figure 2.26(c) Z80 memory write cycle

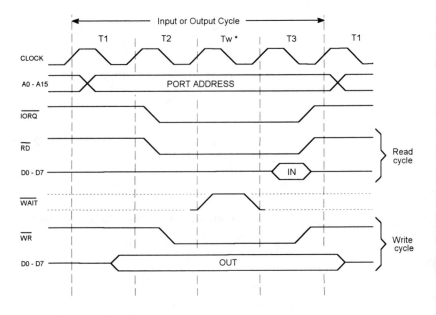

Figure 2.26(d) Z80 input or output cycles

extra cycles, called **wait states**, at a point in the cycle when the bus signals are stable. Wait states are used to slow down the execution rate of an MPU at selected times in order to synchronize its timings with that of certain types of memory or I/O device. A wait state may be inserted by activating the $\overline{\text{WAIT}}$ input of the MPU, or automatic insertion occurs when certain instructions are executed (notably I/O instructions).

Activity

Using the timing diagrams shown in Figure 2.26 (or timing diagrams for any other available MPU), analyse each type of bus cycle. Identify the changes that take place on each clock edge in terms of setting up addresses, generating control signals and performing data transfers. Tabulate your results.

Timing analysis using a CRO

During normal execution of a microcomputer program, waveforms are seldom repetitive. This causes difficulties when attempting to monitor waveforms using a cathode ray oscilloscope (CRO), since this instrument requires repetitive signals. Often a **logic analyser** is used to avoid this problem.

Test loop program

A CRO may still be used if signals can be made repetitive. One solution is to make the MPU execute a small **looping program** that consists of one or two instructions only, thus generating repetitive bus and control signals.

The following Z80 program is an example of such a test loop:

```
START:        LD    A,(1820H)
              JP    START        (START = 1800H)
```

CRO trigger signal

When using internal triggering, a CRO is synchronized to a rising or falling edge of one of the input signals (the slope is user selectable). This may not always give suitable triggering for MPU signals, and does not allow the user to gain any useful reference point.

A better solution is to synchronize the CRO using an externally derived trigger signal, for example program address signals from the address bus. In the Z80 test loop program, A2 provides an acceptable trigger signal, and enables the user to determine the start and finish of the loop sequence (see Figure 2.27).

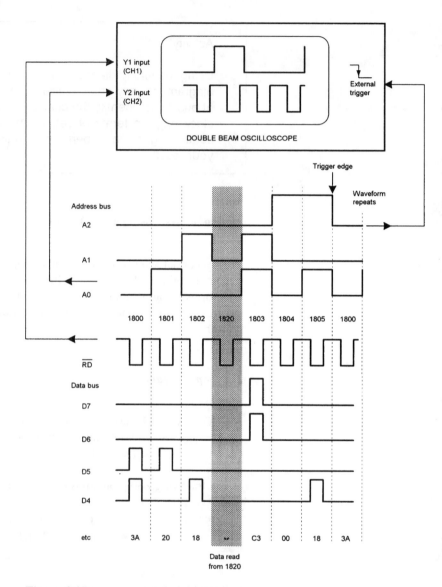

Figure 2.27

Monitoring signals

1   The $\overline{\text{RD}}$ signal provides a useful reference, since all instructions start
with a memory read. Therefore this signal should be displayed on CH1
of the CRO.

2   If the program loop contains any write operations, then this may be
observed simultaneously on CH2 of the CRO.

3   Other signals may be monitored using CH2, and these may include the following:
(a) $\overline{\text{M1}}$
(b) A0, A1, A2, A3
(c) D0 to D7
(d) $\overline{\text{MREQ}}$

Only monitor as many of the address lines as necessary (arrange for all program and data addresses to be closely spaced).

Activity

1   Write a program loop to enable a microcomputer to repeatedly execute a **store accumulator** instruction.
2   Use an oscilloscope to monitor the relevant bus and control signals during execution of your program loop.
3   Draw simple time-related waveforms using information from your CRO, and indicate on the diagrams the activity taking place in relation to your program loop.

Note that due to **queuing** techniques used by some MPUs, e.g. the 8086, an additional known instruction, such as NOP, will be needed immediately outside of the program loop.

Test your knowledge 2.3

1   The first part of an instruction is the:
  A   operator that defines the type of process
  B   operand that defines the type of process
  C   operator that defines the data to process
  D   operand that defines the data to process

2   The second part of an instruction is the:
  A   operator that defines the type of process
  B   operand that defines the type of process
  C   operator that defines the data to process
  D   operand that defines the data to process

3   A microprocessor carries out each instruction by a sequence of operations known as the:
  A   program cycle
  B   read cycle
  C   opcode fetch cycle
  D   fetch–execute cycle

4 Refer to the timing diagram shown in Figure 2.28. Point 'X' on the diagram shows:
A a valid signal on a line
B a valid signal on a bus
C an invalid signal on a line
D an invalid signal on a bus

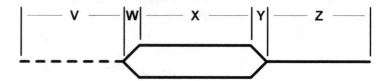

Figure 2.28

5 All instruction cycles start with:
A an input read cycle
B a memory write cycle
C an output write cycle
D an opcode fetch cycle

6 During a memory read cycle, an MPU takes its $\overline{\text{RD}}$ signal:
A high and supplies information to the data bus
B low and supplies information to the data bus
C high and receives information from the data bus
D low and receives information from the data bus

7 During a memory write cycle, an MPU takes its $\overline{\text{WR}}$ signal:
A high and supplies information to the data bus
B low and supplies information to the data bus
C high and receives information from the data bus
D low and receives information from the data bus

8 A microprocessor that makes a distinction between memory read/write cycles and I/O read/write cycles has:
A isolated I/O and uses special I/O instructions
B isolated I/O but does not need special I/O instructions
C memory mapped I/O and uses special I/O instructions
D memory mapped I/O but does not need special I/O instructions

3 Instruction sets and machine code programs

Summary

This element deals with the basic concepts of microprocessor instructions and instruction sets. It covers the different categories of instructions together with their associated addressing modes. Typical instructions for the Z80, 8086 and 8031 microprocessors are analysed. Simple machine code programs are introduced to show how typical instructions are used.

Microprocessor instructions

The function of a microcomputer is to receive data from the outside world, process that data and send the results back to the outside world. Within each microcomputer is a microprocessor (MPU) that is capable of performing a range of simple tasks in response to certain binary inputs. These inputs are called **instructions**, and a logical sequence of instructions, capable of performing complex tasks, is called a **program**.

An MPU requires instructions, in the form of binary numbers, to be presented to its data inputs, one after another in correct program sequence. Binary numbers are too tedious for human use, therefore instructions are normally documented as hexadecimal numbers, or in **mnemonic form** (assembly language), as shown in Figure 3.1.

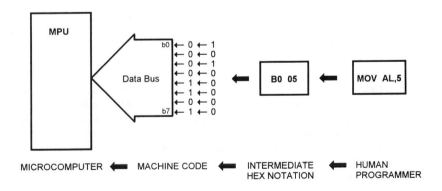

Figure 3.1

The use of assembly language is preferable since it uses instruction mnemonics that are much easier to learn than hex codes, and therefore makes programs much easier to follow.

Instruction format

Each instruction consists of two parts, an operator and an operand, which are written in a manner unique to each MPU, but generally adopting the following format:

> **mnemonic   operand1,operand2**

A **mnemonic** consists of 2 to 6 letters which describe the action of each instruction in an abbreviated form. Mnemonics for the Z80, 8086 and 8031 MPUs are shown in Tables 3.1, 3.2 and 3.3.

Table 3.1   Z80 instruction mnemonics

ADC	*Add operand with carry to accumulator*	LDIR	*As for LDI but repeat until BC = 0*
ADD	*Add operand to accumulator*	NEG	*Negate accumulator (1s complement)*
AND	*Logical AND accumulator with operand*	NOP	*No operation (increment PC)*
BIT	*Test bit in register or memory location*	OR	*Logical OR accumulator with operand*
CALL	*Call subroutine*	OTDR	*Load ouput port(C) with location (HL), decrement HL & B*
CCF	*Complement carry flag*	OTIR	*Load output port (C) with location (HL), increment HL, decrement B*
CP	*Compare operand with acc.*	OUT	*Output to port*
CPD	*Compare location (HL) with acc., decrement HL & BC*	OUTD	*Load output port (C) with location (HL), decrement HL & B*
CPDR	*As for CPD, but repeat until BC = 0*	OUTI	*Load output port (C) with location (HL), increment HL, decrement B*
CPI	*Compare location (HL) with acc., increment HL & BC*	POP	*Recover data from the stack*
CPIR	*AS for CPI, but repeat until BC = 0*	PUSH	*Save data on the stack*
CPL	*Complement accumulator*	RES	*Reset bit in register or memory location*
DAA	*Decimal adjust accumulator*	RET	*Return from subroutine*
DEC	*Decrement register or contents of memory location*	RETI	*Return from maskable interrupt*
DI	*Disable interrupts*	RETN	*Return from non-maskable interrupt*
DJNZ	*Decrement B, jump if not zero*	RL	*Rotate left, through carry, data in register or memory location*
EI	*Enable interrupts*	RLA	*Rotate left, through carry, data in accumulator*
EX	*Exchange registers*	RLC	*Rotate left circular, data in register or memory location*
EXX	*Exchange register sets*	RLCA	*Rotate left circular accumulator*
HALT	*Halt (wait for interrupt or reset)*	RLD	*Rotate digit left and right between acc. and location (HL)*
IM	*Set interrupt mode*	RR	*Rotate right, through carry, data in register or memory location*
IN	*Input from port*	RRA	*Rotate right, through carry, data in accumulator*
INC	*Increment register or contents of memory location*	RRC	*Rotate right circular, data in register or memory location*
IND	*Load location (HL) from port (C), decrement HL & B*	RRCA	*Rotate right circular accumulator*
INDR	*As for IND but repeat until B = 0*	RRD	*Roate digit right and left between acc. and location (HL)*
INI	*Load location (HL) from port (C), increment HL, decrement B*	RST	*Restart to location in operand*
INIR	*As for INI but repeat until B = 0*	SBC	*Subtract operand from accumulator with carry (borrow)*
JP	*Jump to new location*	SCF	*Set carry flag*
JR	*Jump relative to PC*	SET	*Set bit in register or memory location*
LD	*Load register or memory location*	SLA	*Arithmetic shift left of operand*
LDD	*Load location (DE) with location (HL), decrement HL, DE & BC*	SRA	*Arithmetic shift right of operand*
LDDR	*As for LDD but repeat until BC = 0*	SRL	*Logical shift right of operand*
LDI	*Load location (DE) with location (HL), increment DE & HL and decrement BC*	SUB	*Subtract operand from accumulator*
		XOR	*Logical Exclusive-OR accumulator with operand*

Table 3.2 8086 instruction mnemonics

AAA	*ASCII adjust for addition*	JE	*Jump on equal*	LOOPNE	*Loop while not equal*
AAD	*ASCII adjust for division*	JG	*Jump on greater*	LOOPNZ	*Loop while not zero*
AAM	*ASCII adjust for multiplication*	JGE	*Jump on greater or equal*	LOOPZ	*Loop while zero*
AAS	*ASCII adjust for subtraction*	JL	*Jump on less*	MOV	*Move*
ADC	*Add with carry*	JLE	*Jump on less or equal*	MOVS	*Move byte or word (of string)*
ADD	*Add*	JMP	*Jump (within segment)*	MUL	*Multiply*
AND	*Logical AND*	JMPF	*Jump (between segments)*	NEG	*Negate (2s complement)*
CALL	*Call subroutine (within segment)*	JMPS	*Jump (8-bit displacement)*	NOT	*Not (1s complement)*
CALLF	*Call subroutine (between segments)*	JNA	*Jump on not above*	OR	*Logical OR*
CBW	*Convert byte to word*	JNAE	*Jump on not above or equal*	OUT	*Output byte or word to port*
CLC	*Clear carry*	JNB	*Jump on not below*	POP	*Pop data from stack*
CLD	*Clear direction*	JNBE	*Jump on not below or equal*	POPF	*Pop flags from stack*
CLI	*Clear interrupt*	JNC	*Jump on not carry*	PUSH	*Push register onto stack*
CMC	*Complement carry*	JNE	*Jump on not equal*	RCL	*Rotate through carry left*
CMP	*Compare*	JNG	*Jump on not greater*	RCR	*Rotate through carry left*
CMPS	*Compare byte or word (of string)*	JNGE	*Jump on not greater or equal*	REP	*Repeat*
CWD	*Convert word to double word*	JNL	*Jump on not less*	RET	*Return from subroutine (same segment)*
DAA	*Decimal adjust for addition*	JNLE	*Jump on not less or equal*	RETF	*Return from subroutine (between segments)*
DAS	*Decimal adjust for subtraction*	JNO	*Jump on not overflow*	ROL	*Rotate left*
DEC	*Decrement*	JNP	*Jump on not parity*	ROR	*Rotate right*
DIV	*Divide*	JNS	*Jump on not sign*	SAHF	*Store AH into flags*
ESC	*Escape*	JNZ	*Jump on not zero*	SAL	*Shift arithmetic left*
HLT	*Halt*	JO	*Jump on overflow*	SAR	*Shift arithmetic right*
IDIV	*Integer divide*	JP	*Jump on parity*	SBB	*Subtract with borrow*
IMUL	*Integer mulitply*	JPE	*Jump on parity even*	SCAS	*Scan byte or word (of string)*
IN	*Input byte or word from port*	JPO	*Jump on parity odd*	SHL	*Shift left*
INC	*Increment*	JS	*Jump on sign*	SHR	*Shift right*
INT	*Interrupt*	JZ	*Jump on zero*	STC	*Set carry*
INTO	*Interrupt on overflow*	LAHF	*Load AH with flags*	STD	*Set direction*
IRET	*Interrupt return*	LDS	*Load pointer into DS*	STI	*Set interrupt*
JA	*Jump on above*	LEA	*Load effective address*	STOS	*Store byte or word (of string)*
JAE	*Jump on above or equal*	LES	*Load pointer into ES*	SUB	*Subtract*
JB	*Jump on below*	LOCK	*Lock bus*	TEST	*Test*
JBE	*Jump on below or equal*	LODS	*Load byte or word (of string)*	WAIT	*Wait*
JC	*Jump on carry*	LOOP	*Loop*	XCHG	*Exchange*
JCXZ	*Jump on CX zero*	LOOPE	*Loop while equal*	XLAT	*Translate*
				XOR	*Logical Exclusive-OR*

An **operand** identifies data, or the location of data, that will be processed by the instruction, and is separated from the mnemonic by one or more spaces. The Z80, 8086 and 8031 microprocessors refer to operands using the formats shown in Tables 3.4, 3.5 and 3.6.

In principle all instructions require two operands, one to define the *source* of data prior to processing and a second to define the *destination* for processed data. Frequently the source and destination are identical registers or memory locations and need not be separately identified. Many instructions have operand(s) implied in the mnemonic and these also need

Table 3.3   8031 instruction mnemonics

ACALL	*Absolute subroutine call*	LJMP	*Long jump*
ADD	*Add register or byte to acc.*	MOV	*Move data bit, byte or word*
ADDC	*Add register or byte to acc. with carry*	MOVG	*Move code byte*
AJMP	*Absolute jump*	MOVX	*Move external*
ANL	*Logical AND accumulator or carry flag*	MUL	*Mulitply accumulator by register B*
CJNE	*Compare and jump if not equal*	NOP	*No operation*
CLR	*Clear bit or accumulator*	ORL	*Logical OR accumulator or carry flag*
CPL	*Complement bit or accumulator*	POP	*Transfer data byte from stack*
DA	*Decimal adjust (accumulator)*	PUSH	*Transfer data byte to stack*
DEC	*Decrement register or memory*	RET	*Return from subroutine*
DIV	*Divide accumulator by register B*	RETI	*Return from interrupt*
DJNZ	*Decrement register, jump if not zero*	RL	*Rotate accumulator left*
INC	*Increment register, memory or data pointer*	RLC	*Rotate accumulator left through carry flag*
JB	*Jump if bit set*	RR	*Rotate accumulator right*
JBC	*Jump if bit is set and clear bit*	RRC	*Rotate accumulator right through carry flag*
JC	*Jump if carry is set*	SET	*Set bit*
JMP	*Jump indirect*	SJMP	*Short jump*
JNB	*Jump if bit not set*	SUBB	*Subtract with borrow*
JNC	*Jump if carry not set*	SWAP	*Swap nibbles within accumulator*
JNZ	*Jump if accumulator not zero*	XCH	*Exchange accumulator with byte*
JZ	*Jump if accumulator zero*	XCHD	*Exchange digit*
LCALL	*Long call subroutine*	XRL	*Logical Exclusive-OR accumulator or carry flag*

Table 3.4   Z80 instruction operands

Operand	Description	Example
n	8-bit constant in range 0 to 255	LD A,5 (copy constant 5 into the accumulator)
nn	16-bit constant in range 0 to 65535	LD A,(1000) (copy data from location 100H into accumulator)
r,r'	any one of the MPU registers A, B, C, D, E, H, L	LD B,C (copy data from register C into register B)
s	any 8-bit location for all addressing modes allowed for the particular instruction	ADD A,B (add register B to the accumulator)
ss	any 16-bit location for all addressing modes allowed for the particular instruction	INC HL (increment HL register pair)
pp **qq**	any of the register pairs BC, DE, IX, SP any of the register pairs BC, DE, HL, SP	PUSH BC (save BC on the stack)
d	Two's complement displacement in the range +127 to −128	LD A,(IX+4) (copy data from address IX+4 into the accumulator)
e	Extension in relative addressing mode in range −126 to +129	JR 0200 (jump relative to PC to address 0200)

Table 3.5 8086 instruction operands

Operand	Description	Example
reg	a general purpose register of any size	ADD AL,BL or ADD AX,BX
segreg	one of the segment registers: DS, ES, SS or CS (also FS or GS on 80386)	MOV DS,AX
accum	accumulator register of any size: AL or AX (also EAX on the 80386)	ADC AX,5 or AND AL,0F
mem	a direct or indirect memory operand of any size	MOV AL,[2000] or CMP[2000], DX
label	a labelled memory location in the code segment	JMP 2000
src,dest	source or destination memory operand used in a string operation	MOVSB [ES:2000],[1000]
immed	a constant operand of any size	MOV AH,7 or MOV BX,2000

Table 3.6 8031 instruction operands

Operand	Description	Example
Rn	Register R0 to R7 of the selected register bank	e.g. MOV A,R6 (copy register R6 into the accumulator)
direct	8-bit address of internal data location in RAM or SFR	MOV A,7FH (copy data from location 7FH into accumulator)
@Ri	8-bit internal data RAM location addressed indirectly via R0 or R1	MOV A,@R0 (copy data from address R0 into accumulator)
#data	8-bit constant included in instruction	MOV A,#5 (copy constant data 5 into the accumulator)
#data16	16-bit address used by LCAL and LJMP instructions	LJMP 2000H (jump to address 2000H)
addr11	11-bit destination address within same 2K page of program memory	AJMP 0123H (jump to address 0123H)
rel	Two's complement offset which allows relative jumps up to +127 or −128 places from the current address in the PC	JC 0200H (jump on carry to address 0200H)
bit	Directly addressed bit in internal data RAM or SFR (Special Function Registers)	CLR P1.2 (clear bit 2 of Port 1)

not be shown separately. For example the Z80 mnemonic SCF (set carry flag) implies that the carry flag will be used as the operand for a set operation. Therefore the following formats may be encountered.

No separate operands

RET	return from subroutine	(Z80)
RET	return from subroutine	(8086)
RET	return from subroutine	(8031)
data source:	implied	(the stack)
data destination:	implied	(the PC)

This instruction controls program flow by recovering data (*a return address*) from an area of RAM called the **stack** and transferring that data to the program counter.

One operand only

JP	2000H	jump	(Z80)
JMP	2000H	jump	(8086)
LJMP	2000H	long jump	(8031)
data source:		operand	(address 2000H)
data destination:		implied	(the PC)

This instruction copies an address into the program counter and forces program execution to continue from a new address (i.e. causes a program **jump**).

Identical source and destination operands

INC	A	increment accumulator	(Z80)
INC	AL	increment accumulator AL	(8086)
INC	A	increment accumulator	(8031)
data source:		operand	(the accumulator)
data destination:		operand	(the accumulator)

This instruction increments (adds 1) to the accumulator. Effectively the source data is that contained in the accumulator, and after processing, the destination is the accumulator.

Two operands

```
LD    A,B        load accumulator with register B    (Z80)
MOV  AL,BL      move BL into AL                      (8086)
MOV  A,R1       move R1 into A                       (8031)


data source:              operand 2      (register B, BL or R1)
data destination:         operand 1      (the accumulator)
```

This instruction copies the source operand into the destination operand (A or AL). Note that all LD/MOV instructions have both source and destination operands.

Addressing modes

An MPU may access data in registers, memory or I/O using a number of different ways to identify the source or destination of that data. Each method of identifying the location of data is known as an **addressing mode**, and depending upon their architecture, different types of MPU have their own particular addressing modes. The following addressing modes are commonly used.

Implied

The operand is implied in the operator part of the instruction.

```
SCF    set carry flag          (Z80)
STC    set carry flag          (8086)
RET    return from subroutine   (8031)
```

The carry flag is implied in the first two instructions (the 8031 equivalent does not use implied addressing, therefore the RET instruction is included in which the stack and program counter are implied).

Immediate

The operand is a **constant** that forms part of the program code, and is stored in memory location(s) that immediately follow the one(s) in which the opcode is stored (see Figure 3.2).

```
LD    A,5        load accumulator with 5              (Z80)
MOV  AL,5       move data 5 into accumulator AL       (8086)
MOV  A,#5       move byte 5 into A                    (8031)
```

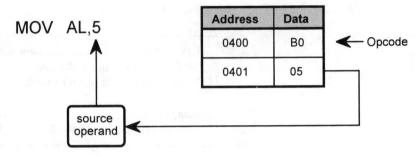

Figure 3.2

In each of these instructions, constant data 5 is copied into the accumulator. This data is *embedded* in the program as the last byte of the instruction code.

Register

The operand is specified as the contents of one of the internal registers of the MPU.

```
ADD  A,B      add register B to accumulator        (Z80)
ADD  AL,BL    add register BL to accumulator AL     (8086)
ADD  A,R0     add register R0 to accumulator        (8031)
```

The source operand is given as the contents of register B, BL or R0, and this is added to the contents of the accumulator (A or AL).

Direct

This is sometimes called **extended** or **absolute** addressing in which the data bytes that follow the opcode form an address that defines the *location* of the data to be used by the instruction (see Figure 3.3).

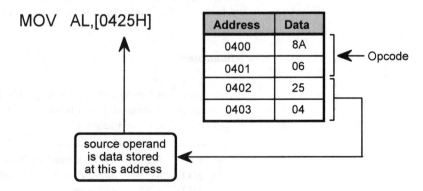

Figure 3.3

LD	A,(0425H)	load accumulator with data at 0425H	(Z80)
MOV	AL,[0425H]	move data at 0425H to accumulator AL	(8086)
MOV	A,7FH	move byte at 7FH to accumulator	(8031)

Data stored at an address specified by the operand is copied into the accumulator. Note that in the case of the 8031 only an 8-bit address may be specified.

Indirect

This is similar to memory addressing, except that the indirect address is defined as the contents of a 16-bit register (see Figure 3.4).

LD	HL,0425H	load pointer register	(Z80)
LD	A,(HL)	load accumulator from 0425H	(Z80)
MOV	BX,0425H	load pointer register	(8086)
MOV	AL,[BX]	move data from 0425H into AL	(8086)
MOV	DPTR,#0425H	load pointer register	(8031)
MOVX	A,@DPTR	move byte from 0425H into A	(8031)

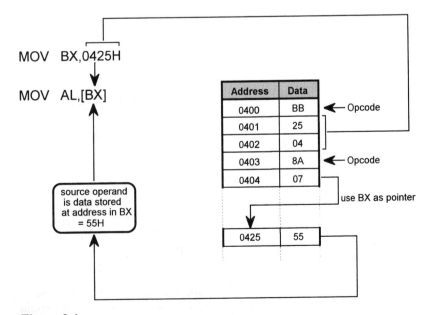

Figure 3.4

In all cases shown, a 16-bit register is loaded with an address of data (0425H), and the register is sometimes called a **pointer** since it is considered to point to the data at that address. Each of the above instructions then copies data stored at the pointer location into the accumulator.

Addressing mode combinations

Source and destination operands may each have their own separate addressing modes, therefore an instruction may use different addressing mode combinations as shown in the following examples.

Example 1

```
ADD    A,(HL)      (Z80)
ADD    AL,[BX]     (8086)
ADD    A,@R0       (8031)
```

The instructions in this example all use indirect addressing to specify the data source, but use register addressing to specify the location of the data to which the source byte must be added.

Example 2

```
LD     A,05        (Z80)
MOV    AL,05       (8086)
MOV    A,#05       (8031)
```

The instructions in this example all use immediate addressing to specify the source of data to transfer, but use register addressing to specify its destination.

Example 3

```
JP     2000H       (Z80)
JMP    2000H       (8086)
LJMP   2000H       (8031)
```

The instructions in this example all use immediate addressing for the source operand (2000H), but implied addressing for the destination, the PC or program counter (or IP, instruction pointer for the 8086).

Test your knowledge 3.1

1 The Z80 instruction LD (1900H),A copies:
A   data from A into memory location 1900H
B   hex number 19H into A
C   data from memory location 1900H into A
D   hex number 00H into A

2 If register B contains 05 and register C contains 03,
then after execution of the Z80 instruction LD B,C
register B contains:
A   03 and C contains 05
B   05 and C contains 03
C   03 and C contains 00
D   03 and C contains 03

3 If accumulator A contains 25H, HL contains 1900H and
the data stored at address 1900H is 33H, then after
execution of the Z80 instruction LD A,(HL) the
accumulator A will contain:
A   19H
B   33H
C   25H
D   00H

4 The 8086 instruction MOV AL,[1020H] copies:
A   hex number 20H into AL
B   data from memory location 1020H into AL
C   data from AL into memory location 1020H
D   hex number 10H into AL

5 If register BL contains 05 and register CL contains
03, then after execution of the 8086 instruction
MOV BL,CL register BL contains:
A   03 and CL contains 00
B   03 and CL contains 03
C   03 and CL contains 05
D   05 and CL contains 03

6 If accumulator AL contains 25H, BX contains 1900H
and the data stored at address 1900H is 33H, then after
execution of   the 8086 instruction MOV AL,[BX]
accumulator AL will contain:
A   19H
B   00H
C   25H
D   33H

7 The 8031 instruction MOV A,7FH copies:
   A hex number 7FH into accumulator A
   B accumulator A into memory location 7FH
   C data from address 7FH into accumulator A
   D data from accumulator A into address 7FH

8 If register R0 contains 05 and accumulator A contains 03, then after execution of the 8031 instruction MOV A,R0 accumulator A contains:
   A 03 and register R0 contains 00H
   B 05 and register R0 contains 03H
   C 03H and register R0 contains 03H
   D 05H and register R0 contains 05H

9 If accumulator A contains 25H, register R0 contains 7FH and the data stored at address 7FH is 33H, then after execution of the 8031 instruction MOV A,@R0 accumulator A will contain:
   A 25H
   B 33H
   C 7FH
   D 00H

Activity

1 Obtain instructions in the use of microprocessor debugging programs for one or more of the microprocessors being considered so that you are able to carry out the following operations:

   (a) enter programs using machine code or assembly language,

   (b) inspect/alter the program data in memory,

   (c) inspect/alter the microprocessor registers,

   (d) trace program execution one instruction at a time.

2 Check your answers to the 'test your knowledge' problems by first setting initial values of registers and memory locations as necessary, then executing the relevant instruction.

3 Check each instruction with different data until you are sure that you understand its operation, then record your results.

Information provided by an instruction set

An instruction set provides information concerning each instruction which may be of use to a programmer, including the following items.

Mnemonics

Abbreviations representing every instruction in the set, identifying instructions available with a particular MPU, and showing their correct assembly language syntax.

Symbolic notation

A shorthand method of representing the effect of executing an instruction.

Opcodes

The actual machine codes for each instruction, sometimes useful for testing or modifying a program (*patching*), and essential when carrying out hardware testing.

Flags

Shows which flags are affected and how they change in response to the execution of each instruction – important for selection processes within a program (*decision branches*).

Number of bytes

This information allows a programmer to determine the length of a program, or part of a program, for memory allocation purposes.

T states

This gives the number of clock cycles required to execute each instruction and allows a programmer to determine the total *execution time* of a program.

Categories of instructions

Instructions may be regarded as falling into one of the following three categories.

Data transfer instructions

These instructions transfer data between registers, memory locations or I/O ports, and with the exception of exchange instructions, are **copy operations** which do not change the source data. Execution of most data transfer instructions therefore leaves both source and destination with identical data. The most common data transfer instructions are **load**, **move** and **store**, all of which may be considered as copy operations. Most MPUs have specialized stack-related transfers such as **push** and **pop**, and some MPUs have instructions to exchange data between registers or between each half of a register, but few MPUs are able to transfer data directly between external memory locations.

The following examples show typical data transfer instructions.

(a) Z80

LD	A,5	*transfer constant data 5 into the accumulator*
OUT	0,A	*transfer data in A to output Port 0*
PUSH	BC	*transfer the contents of register pair BC to the stack*

(b) 8086

MOV	AL,5	*transfer constant data 5 into accumulator AL*
OUT	0	*transfer data in AL to output Port 0*
PUSH	BX	*transfer contents of BX (BH and BL) to the stack*

(c) 8031

MOV	A,#5	*transfer constant data 5 into the accumulator*
MOV	A,@R0	*transfer data from address in R0 to the accumulator*
PUSH	20H	*transfer data from address 20H to the stack*

Example 4

Write a Z80 instruction sequence (program) to copy the data in memory location 1900H into 1901H using indirect addressing.

```
LD      HL,1900    ; HL is a pointer to address 1900H
LD      A,(HL)     ; get data stored at 1900H into accumulator
INC     HL         ; HL now points to address 1901H
LD      (HL),A     ; copy data in accumulator into address 1901H
                   ; add suitable terminating instruction
```

Example 5

Write an 8086 instruction sequence (program) to copy the data in memory location 1900H into 1901H using indirect addressing.

```
MOV     BX,1900    ; BX is a pointer to address 1900H
MOV     AL,[BX]    ; get data stored at 1900H into accumulator AL
INC     BX         ; BX now points to address 1901H
MOV     [BX],AL    ; copy data in AL into address 1901H
                   ; add suitable terminating instruction
```

Example 6

Write an 8031 instruction sequence (program) to copy the data in memory location 1900H into 1901H using indirect addressing.

```
MOV     DPTR,#1900 ; DPTR is a pointer to address 1900H
MOVX    A,@DPTR    ; get data stored at 1900H into accumulator
INC     DPTR       ; DPTR now points to address 1901H
MOVX    @DPTR,A    ; copy data in A into address 1901H
                   ; add suitable terminating instruction
```

Arithmetic and logic

These instructions perform the basic arithmetic and logic operations within a microprocessor, and all MPUs perform the following arithmetic and logic operations:

Arithmetic: add, subtract, increment, decrement and shift
Logic: OR, AND, Exclusive-OR, shift and rotate

Some MPUs also have instructions to perform *multiply* and *divide*, but if not available, these functions may be implemented by small programs using basic arithmetic and logic functions. All arithmetic operations are carried out using two's complement arithmetic, and may include the carry flag in their calculations.

Logic instructions operate between **corresponding bits** of the **two operands**. For example, if a logical AND instruction is executed, then bit 0 of operand 1 will be ANDed with bit 0 of operand 2, therefore bit 0 of the result will only be at logical 1 if both corresponding bits of the operands are at logical 1.

Typical arithmetic and logic instructions are:

(a) Z80

ADD	A,5	*add constant 5 to the accumulator*
SUB	5	*subtract constant 5 from the accumulator*
OR	0FH	*bitwise logical OR of accumulator with constant 0FH*
AND	0FH	*bitwise logical AND of accumulator with constant 0FH*
XOR	0FH	*bitwise logical XOR of accumulator with constant 0FH*

Note that the Z80 accumulator must contain one of the operands and the result is always stored back into the accumulator.

(b) 8086

ADD	AL,5	*add constant 5 to accumulator AL*
SUB	AL,5	*subtract constant 5 from accumulator AL*
OR	AL,0FH	*bitwise logical OR of AL with constant 0FH*
AND	AL,0FH	*bitwise logical AND of AL with constant 0FH*
XOR	AL,0FH	*bitwise logical XOR of AL with constant 0FH*

These examples all use accumulator AL, but any general purpose register can act as a destination operand, e.g. ADD BL,5 or AND CL,0FH.

(c) 8031

ADD	A,#5	*add constant 5 to accumulator*
SUBB	A,#5	*subtract constant 5 with borrow from accumulator*
ORL	A,#0FH	*bitwise logical OR of accumulator with constant 0FH*
ANL	AL,#0FH	*bitwise logical AND of accumulator with constant 0FH*
XRL	AL,#0FH	*bitwise logical XOR of accumulator with constant 0FH*

Note that the 8031 MPU does not have an instruction to subtract without inclusion of the borrow (carry) flag, therefore it is the responsibility of the programmer to ensure the carry flag is of known state prior to use of this instruction.

Example 7

Write a Z80 instruction sequence (program), using direct (extended) addressing, to fetch data from address 1900H into the accumulator, add 5 to it and store the sum in address 1901H.

```
LD    A,(1900)     ; get data from address 1900H into accumulator
ADD   A,5          ; add 5 to accumulator
LD    (1901),A     ; store the sum at address 1901H
                   ; add suitable terminating instruction
```

Example 8

Write an 8086 instruction sequence (program), using direct (extended) addressing, to fetch data from address 1900H into the accumulator, add 5 to it and store the sum in address 1901H.

```
MOV   AL,[1900]    ; get data from address 1900H into AL
ADD   AL,5         ; add 5 to accumulator AL
MOV   [1901],AL    ; store the sum at address 1901H
                   ; add suitable terminating instruction
```

Example 9

Write an 8031 instruction sequence (program), using direct addressing, to fetch data from address 70H into the accumulator, add 5 to it and store the sum in address 71H.

```
MOV   A,70         ; get data from address 70H into accumulator
ADD   A,#5         ; add 5 to accumulator
MOV   71,A         ; store the sum at address 71H
                   ; add suitable terminating instruction
```

Program control (test and branch)

Program instructions are executed strictly in **stored order**, using addresses supplied by the PC (program counter). This continues until the natural progression of the PC is changed, thus forcing execution to continue from a different place in the program, i.e. a **jump** or **branch** is implemented. This action may occur at any point in a program, and may be conditional upon the result of a test, thus causing selective execution of sections of a program and providing its basic intelligence. Instructions in this category provide the means of implementing program branches, and typical examples are:

(a) Z80

JP	Z,0200H	*jump to address 0200H if the Z flag is set*
JP	NZ,0200H	*jump to address 0200H if the Z flag is clear*
JP	C,0200H	*jump to address 0200H if the C flag is set*
JP	NC,0200H	*jump to address 0200H if the C flag is clear*
DJNZ	0200H	*decrement register B, jump to 0200H if B is not zero*

(b) 8086

JZ	0200H	*jump to address 0200H if ZF is set*
JNZ	0200H	*jump to address 0200H if ZF is clear*
JC	0200H	*jump to address 0200H if CF is set*
JNC	0200H	*jump to address 0200H if CF is clear*
LOOP	0200H	*decrement CX, jump to 0200H if CX not zero*

Note that the 8086 provides up to **three mnemonics** for the **same instruction**. Therefore JC (jump if carry), JB (jump if borrow) and JNAE (jump if not above or equal) all have the same instruction opcode, all test the state of the carry flag, and cause a jump if CF = 1.

(c) 8031

JZ	0200H	*jump to address 0200H if accumulator is zero*
JNZ	0200H	*jump to address 0200H if accumulator not zero*
JC	0200H	*jump to address 0200H if CY flag is set*
JNC	0200H	*jump to 0200H if CY flag is clear*
DJNZ	R2,0200H	*decrement register R2, jump to 0200H if R2 not zero*

Note that the 8031 does not have a zero flag, and JZ and JNZ instructions test the accumulator for zero.

Example 10

Write a Z80 instruction sequence (program) to add 5 to the accumulator three times using program control (loop) instructions.

```
        LD    B,3      ; counter to execute loop three times
        LD    A,0      ; start off with zero in the accumulator
SUM:    ADD   A,5      ; add 5 to accumulator
        DJNZ  SUM      ; repeat addition until count in B is zero
                       ; add suitable terminating instruction
```

Note: SUM is the address at which the ADD A,5 instruction is located.

Example 11

Write an 8086 instruction sequence (program) to add 5 to accumulator AL three times using program control (loop) instructions.

```
        MOV   CX,3     ; counter to execute loop three times
        MOV   AL,0     ; start off with zero in the accumulator
SUM:    ADD   AL,5     ; add 5 to accumulator
        LOOP  SUM      ; repeat addition until CX is zero
                       ; add suitable terminating instruction
```

Note: SUM is the address at which the ADD AL,5 instruction is located.

Example 12

Write an 8031 instruction sequence (program) to add 5 to the accumulator three times using program control (loop) instructions.

```
        MOV   R0,#3    ; counter to execute loop three times
        CLR   A        ; start off with zero in the accumulator
SUM:    ADD   A,#5     ; add 5 to accumulator
        DJNZ  R0,SUM   ; repeat addition until R0 is zero
                       ; add suitable terminating instruction
```

Note: SUM is the address at which the ADD A,#5 instruction is located.

Activity

1   Obtain instructions in the use of microprocessor
    debugging programs for one or more of the micro-
    processors being considered so that you are able to
    carry out the following operations:
    (a) **enter programs** using machine code or
        assembly language,

(b) **inspect/alter** the program data in **memory**,
(c) **inspect/alter** the microprocessor **registers**,
(d) **trace program** execution one instruction at a time.

2 Using an appropriate address in RAM, enter the examples given for each category, step through them to check that they behave as expected and record your results.

Test your knowledge 3.2

1 The Z80 instruction LD (1900H),A belongs to the following category of instructions:
   A arithmetic
   B logic
   C data transfer
   D program control

2 The Z80 instruction JR NC,0200H belongs to the following category of instructions:
   A arithmetic
   B logic
   C data transfer
   D program control

3 The Z80 instruction XOR B belongs to the following category of instructions:
   A arithmetic
   B logic
   C data transfer
   D program control

4 The 8086 instruction LOOP 2000H belongs to the following category of instructions:
   A arithmetic
   B logic
   C data transfer
   D program control

5 The 8086 instruction ADC AL,5 belongs to the following category of instructions:
   A arithmetic
   B logic
   C data transfer
   D program control

6   The 8086 instruction MOV AL,AH belongs to the
    following category of instructions:
    A   arithmetic
    B   logic
    C   data transfer
    D   program control

7   The 8031 instruction MOV A,7FH belongs to the
    following category of instructions:
    A   arithmetic
    B   logic
    C   data transfer
    D   program control

8   The 8031 instruction ANL A,R0 belongs to the
    following category of instructions:
    A   arithmetic
    B   logic
    C   data transfer
    D   program control

9   The 8031 instruction LJMP 2000H belongs to the
    following category of instructions:
    A   arithmetic
    B   logic
    C   data transfer
    D   program control

Register summary

Z80

The Z80 MPU has 18 8-bit registers and four 16-bit registers that are directly
accessible to the programmer. The registers include two sets of six general
purpose registers that may be used individually as 8-bit data registers or in
pairs as 16-bit address or data registers. There are two accumulators and flag
registers and six special purpose registers, as shown in Figure 3.5.

8086

The 8086 family of MPUs share the same set of 14 16-bit registers, four of
which may be accessed as eight 8-bit registers. In the 80386 upwards, most
registers can also be accessed as extended 32-bit registers. The registers
directly accessible to the programmer are shown in Figure 3.6.

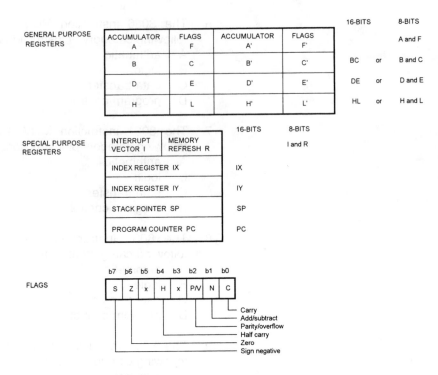

Figure 3.5 Z80 register set

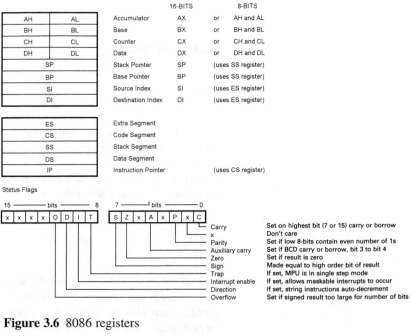

Figure 3.6 8086 registers

8031

The 8031 MPU is intended for use in industrial control applications rather than desktop computers, therefore its register arrangements are somewhat different to those of the previous Z80 and 8086 MPUs. The 8031 has 128 bytes of internal RAM (256 bytes in 8032/8052 devices), and the lowest 32 bytes are grouped into four banks of eight general purpose registers, **R0** through **R7**. Two bits in the **Program Status Word** (PSW) select which bank is in use. There are 16-bit registers and also **Special Function Registers** (SFR) occupying a further 128 bytes of RAM. Since registers are implemented in internal RAM they may also be accessed in terms of their memory addresses. Registers accessible to the programmer are shown in Figure 3.7.

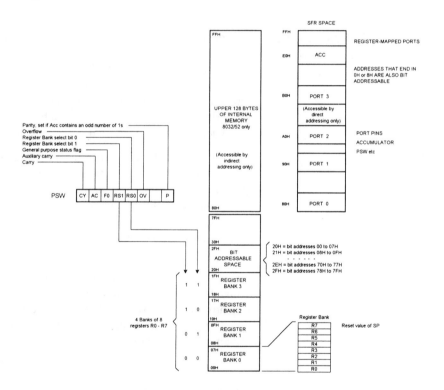

Figure 3.7 8031 registers

Registers and certain internal memory locations are also **bit addressable**, thus ACC.2 addresses bit 2 of the accumulator. The 8031 also contains a single bit or 'Boolean' processor which uses the C flag as a single bit accumulator for certain operations. This facilitates logic operations which involve processing individual pairs of bits.

4 Memory systems

Summary

This element deals with different types of memory devices, their characteristics, and shows how they are connected into microcomputer circuits. Bus structures and the related use of tri-state buffers are also covered, as is the need for address decoding. Practical circuit examples are shown.

Microcomputer memories

A microprocessor performs its tasks by executing a sequence of stored instructions called a program. It therefore follows that all microcomputers must provide **storage facilities** for program instructions and also for variable data being processed by their MPUs. Simple control systems require much less memory than is the case for desktop computers, otherwise, in principle, there is little difference between the two systems.

Microcomputer memory devices must provide the following facilities:

1 A number of individual **storage elements** known as memory cells, each capable of temporary or permanent storage of a single binary digit.
2 A **system of addressing** which provides a means of selecting a specific memory cell (or group of cells) within the memory device.
3 A means of **writing data** into a specific memory location.
4 A means of **reading data** from a specific memory location.

Memory devices may be classified as **volatile** or **non-volatile**. A volatile memory loses its stored data if the power supply to it is not continuously maintained. A non-volatile memory retains its stored data permanently, and removal of its power supply does not result in loss of data.

The following types of memory device are commonly found in microcomputer systems.

Random access memory

This is a type of memory in which any particular storage location (address) may be directly selected without first having to sequence through other locations, i.e. one storage location may be accessed as easily as any other regardless of its physical position within the memory device. Any type of memory may randomly access data, although the term RAM has generally become associated with **read-write memory**, i.e. memory in which the user may either read from or write to specified memory locations while still in its normal circuit environment. Programs stored in RAM are generally referred to as software.

Two different types of read-write memory are used in microcomputer systems. Both types are volatile, and these are:

1 **Static RAM (SRAM)** in which bistable or flip-flop circuits are used as storage elements such that when a bistable is set, it stores a logical 1, and when it is reset it stores a logical 0 (or vice versa). With this type of memory, unless overwritten, the stored data is retained permanently provided the power supply is continuously maintained to keep the bistables in their current states. After interruption of the power supply the bistables have unpredictable states and the memory is filled with random data, i.e. original data is lost. The block diagram and pin out of a typical static RAM is shown in Figures 4.1(a) and (b).

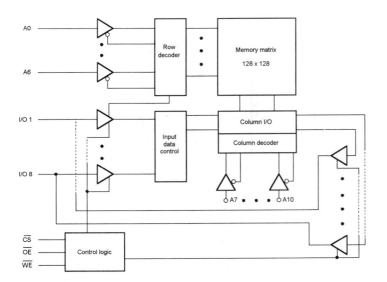

Figure 4.1(a)

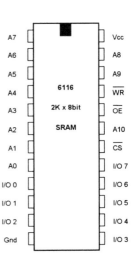

Figure 4.1(b)

The use of bistables or flip-flops means that each memory cell consists of a relatively large number of components, therefore there are practical limits to the size of a static RAM that can be manufactured. Also the power consumption is relatively high, even for CMOS types of RAM, since each memory cell has at least one transistor passing current.

2 **Dynamic RAM (DRAM)** in which small capacitors are used as temporary storage elements such that a charged capacitor is considered to store a logical 1 and when discharged, stores a logical 0. The block diagram and pin out of a dynamic RAM is shown in Figures 4.2(a) and (b). These are often manufactured into memory module units as shown in Figure 4.2(c).

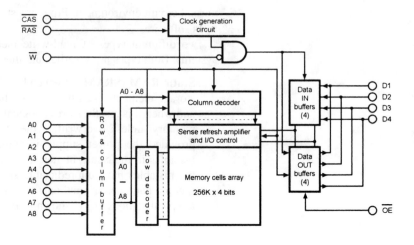

Figure 4.2(a) DRAM Block diagram

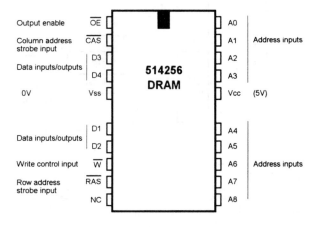

Figure 4.2(b) 256K x 4-bit DRAM pin out diagram

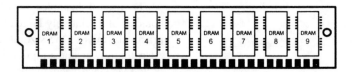

Figure 4.2(c) 4M x 9-bit SIMM (Single In-line Memory Module)

Leakage effects invariably cause loss of charge and hence loss of stored data, therefore the charges on these capacitors must be restored at frequent intervals (approximately every 2 ms) to prevent loss of data, and this is a process known as **refreshing**. Interruption of the power supply also causes loss of data since the capacitors become discharged.

Because each memory cell contains few components, the packing density is high and it is possible to manufacture dynamic RAMs with very large storage capacity but small physical size. Also, since current flow is required only when charging or discharging the internal capacitors, current consumption is very low, and dynamic RAMs are the only practical devices for very large microcomputer memories.

Read-only memory (ROM)

This is non-volatile memory that is used for the storage of permanent non-alterable data. Data is stored in ROM during the final metallizing stages of manufacture by means of a **mask programming** technique, using a mask constructed according to bit patterns supplied by the customer. The construction of such a mask is relatively expensive, therefore this type of device is used mainly for applications where high volume production is anticipated. Once manufactured, a ROM cannot be reprogrammed for a different application. The block diagram and pin outs of a typical ROM are shown in Figures 4.3(a) and (b). Note that active levels for chip select inputs may also be specified by the customer.

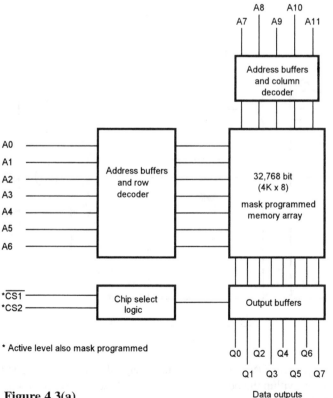

Figure 4.3(a)

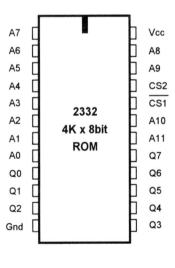

Figure 4.3(b)

Programmable read-only memory (PROM)

A PROM behaves in an identical manner to a ROM except that programming does not take place during manufacture, but instead is programmed by the user, i.e. it is **field programmable**. A PROM is supplied by its manufacturer in a blank state, with all bits held at the same logic level by means of fusible links. Programming a PROM involves the user blowing (or fusing) links in selected locations, thus changing the logic level in these locations only. Like a ROM, a PROM cannot be reprogrammed for a different application, although it is sometimes possible to make minor changes, but only if it entails blowing additional links (links blown already cannot be restored).

Erasable programmable read-only memory (EPROM)

An EPROM performs an identical function to a ROM, and like a PROM it is also field programmable, but with the additional advantage that it may be erased and reprogrammed. An EPROM uses floating gate avalanche (FAMOS) technology, and is supplied in a blank state by its manufacturer. The block diagram of a typical EPROM is shown in Figure 4.4.

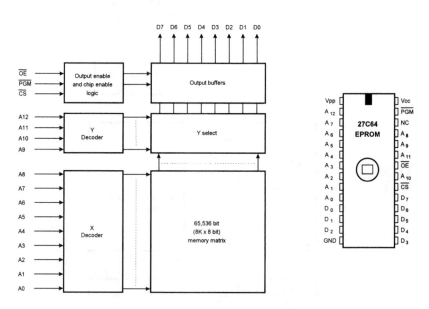

Figure 4.4

Programming an EPROM involves storing electrical charges on selected FET floating gates within the device, thus changing these memory locations to the opposite logic state. This is usually carried out using a special piece

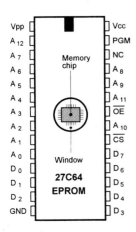

Figure 4.5

of equipment called an **EPROM programmer**. The charges are held by the floating gates for a period of at least 10 years, therefore the data may be considered as permanently stored. However, an EPROM may be erased at any time by removing it from circuit, followed by exposure to ultraviolet light (wavelength 2537 Å) for 10 to 20 minutes. A small window is built into an EPROM specifically for this purpose (see Figure 4.5).

Unfortunately this process erases the entire memory and selected locations cannot be erased and reprogrammed, although as with PROMs, minor changes can sometimes be effected if they are in the right direction. One time programmable (OTP) devices are also available which are cheaper than ordinary EPROMs, but which do not have a window fitted for erasure. These devices are used in systems where reprogramming is not required, but in which the cost of a mask-programmed ROM cannot be justified.

Electrically erasable read-only memory (EEPROM)

An EEPROM performs the same basic functions as an EPROM and it also uses the principle of trapped charges for data storage, but after programming, data may be changed electrically by application of suitable control signals. Unlike an EPROM, an EEPROM does not have to be completely erased before it can be reprogrammed, nor does it have to be removed from circuit, but selected blocks of memory or even individual memory locations may be erased while in situ (see Figures 4.6(a) and (b)).

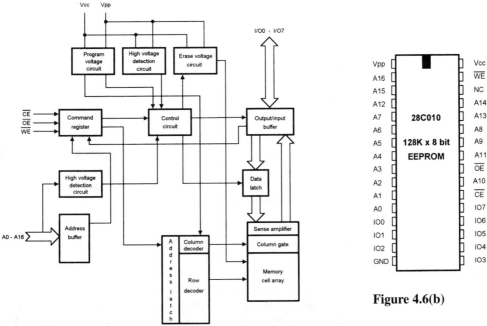

Figure 4.6(b)

Figure 4.6(a)

Certain types of programs are stored in ROM, PROM, EPROM or EEPROM and these are then known as **firmware.**

Memory characteristics

The following characteristics may be considered when selecting a memory device for a microcomputer.

Number of locations

The number of address lines on a memory is an indication of the number of different locations (or addresses) available within the device. If a memory has 'n' address lines, then it provides 2^n different storage locations. Since the lowest order address line of a memory is A0, the total number of address lines is equivalent to the number of the highest address line plus one, i.e. if the highest address line is A11, then there are 12 address lines. The number of storage locations in such a memory is 2^{12} or 4096_{10} (4K). Other sizes of memory may be determined by reference to Table 4.1.

Table 4.1

Address bus lines	Number of address lines	Number of addresses	Size of memory
A0–A9	10	1 024	1K
A0–A10	11	2 048	2K
A0–A11	12	4 096	4K
A0–A12	13	8 192	8K
A0–A13	14	16 384	16K
A0–A14	15	32 768	32K
A0–A15	16	65 536	64K
A0–A16	17	131 072	128K
A0–A17	18	262 144	256K
A0–A18	19	524 288	512K
A0–A19	20	1 048 576	1M

Number of bits

Each address must store at least one bit, but it is possible for addresses to store a number of bits, e.g. 4, 8 or 16 bits. This means that when a particular address is applied to the memory device, a number of independent bits are simultaneously available (i.e. parallel bits), thus enabling it to associate complete instructions, characters or numeric data at a single address. The number of data I/O lines on a memory indicates how many bits are stored at each address.

Storage capacity

The total storage capacity of a memory device is often quoted in manufacturer's literature, and is the product of storage locations and number of bits, i.e.

$$\text{Total storage capacity} = \text{number of locations} \times \text{number of bits at each location}$$

Type of access

Memory access may be **random** or **sequential** (serial). Most microcomputer memory devices use random access, although sequential access memories may be found in EEPROMs used with certain types of microcontroller. Random access has already been defined as a means of directly accessing required data without having first to sequence through other storage locations, i.e. an address is used to select the required data. When using sequential access, however, a particular location cannot be accessed directly, but data must be brought out of memory in its stored sequence until the required data is reached.

Access time

This is the time taken for a memory device to deliver data after receiving valid address and control signals. In older types of memory this could be as long as 1000 ns, but is now generally 100 ns or less for random access MOS devices. The access time for desktop computer DRAMs is typically 70 ns. Access times for sequential memories are generally longer, and vary according to the location of data within the device.

Volatility

The ability of a memory to retain its stored data after removal of the power supply to a microcomputer is a measure of its volatility. In general, static and dynamic RAMs are volatile and therefore lose their stored data after removal of the power supply, whereas all types of ROM are non-volatile and retain their stored data. Sometimes it is essential that variable data is retained after loss of the power supply (e.g. '*set up*' information for a personal computer), and for such applications non-volatile RAM is required. Such storage may be provided by battery backup, in which a small rechargeable battery is used to prevent volatile RAM from losing data, or by the use of special non-volatile RAMs or even EEPROMs.

Dynamic or static

It has already been stated that RAM may be either static or dynamic. Static RAM is simple to use, but the cost per bit is higher than for dynamic RAM and in large memory systems the power consumption is unacceptably high. Dynamic RAM has lower cost per bit and much lower power consumption, but does require the added cost and complication of refreshing. Therefore in small microcontroller systems with a few kilobytes of memory, static RAM is normally used, but in desktop computers with several megabytes of memory, dynamic RAM is used.

Test your knowledge 4.1

1   The storage method used in a static RAM is:
A   mask programming
B   capacitance
C   fusible links
D   bistables

2   The storage method used in a dynamic RAM is:
A   mask programming
B   capacitance
C   fusible links
D   bistables

3   A volatile memory that requires refreshing every 2 ms is called:
A   PROM
B   static RAM
C   ROM
D   dynamic RAM

4   A RAM is:
A   non-volatile read-write memory
B   non volatile read-only memory
C   volatile read-write memory
D   volatile read-only memory

5   Non-volatile memory that may be erased and re-programmed is called:
A   SRAM
B   DRAM

C   PROM

D   EPROM

6   The number of different addresses in a memory IC
     with pins marked A0 to A10 is:
     A   10
     B   1K
     C   2K
     D   11

7   A memory device with 12 address lines and 8 data
     outputs has a total storage capacity of:
     A   12 bytes
     B   4K bits
     C   12 bits
     D   32K bits

8   The access time for a typical RAM is:
     A   100 ns
     B   1000 ns
     C   100 μs
     D   1000 μs

Bus structure

In order to simplify the wiring of a microcomputer, a bus structure is used in which parallel groups of conductors, called **buses**, are used to convey information from one part of the system to another. All major components of a microcomputer are connected between the following three buses:

1   **Data bus**,
2   **Address bus**, and
3   **Control bus**

When using a bus structure, only one device (*a talker*) is permitted to transfer information to the data bus at any given time, otherwise a situation may arise where two devices simultaneously try to drive a data line to opposite logic levels. This situation is called a **bus conflict**, or **bus contention**, and leads to incorrect operation and possible damage to components.

A less serious, but equally undesirable situation occurs when two inputs (*listeners*) are actively connected to a data line at the same time. Only one device should receive information from the data bus at any instant.

The MPU is either a talker or a listener for all data transfers, since information cannot be transferred directly between memory or I/O locations. Memory to memory data transfers must take place via the MPU.

Except for systems that use direct memory access (DMA), all addresses are generated by the MPU, therefore it is the only talker on the address bus, and under normal circumstances no logical conflicts should occur.

Tri-state buffers

A buffer is a device that is connected between two parts of a system to prevent **unwanted interaction**, and frequently consists of a **current amplifier** or a **small memory**. Without the use of a buffer, many parts of a system would interfere with one another to such an extent that correct operation could not be achieved. A conventional buffer used in a logic circuit would have its output at one of two different logic states, logical 0 (low) or logical 1 (high), but for bus systems, **tri-state buffers** are used which have a third output state of **high impedance** (open circuit). An additional input is provided on a tri-state buffer for enabling or disabling its output. When enabled, the buffer output may be logical 0 or logical 1 depending upon the state of its data input, but when disabled the buffer behaves as though its output had been disconnected from the circuit. Different types of tri-state buffer are available, and the operation of each type may be studied by reference to Figure 4.7.

(a) Non-inverting, active low

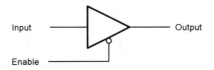

Enable	Input	Output
0	0	0
0	1	1
1	0	High Impedance
1	1	High Impedance

(b) Inverting, active low

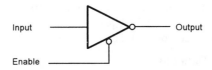

Enable	Input	Output
0	0	1
0	1	0
1	0	High Impedance
1	1	High Impedance

(c) Non-inverting, active high

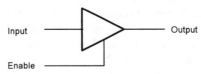

Enable	Input	Output
0	0	High Impedance
0	1	High Impedance
1	0	0
1	1	1

(d) Inverting, active high

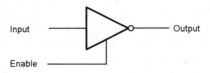

Enable	Input	Output
0	0	High Impedance
0	1	High Impedance
1	0	1
1	1	0

Figure 4.7

The normal method of avoiding bus conflicts in a bus structured system is to use memory and I/O devices with tri-state buffers so that they are physically connected to the system buses, but are not actively connected until enabled or selected by means of appropriate signals applied to their **chip enable** (CE) or **chip select** (CS) inputs. Therefore ROM, RAM and input devices can all act as talkers on a common data bus without the danger of a bus conflict. When not selected, the data I/O lines of tri-state devices effectively become open circuit or high impedance, and electrically may be considered as disconnected from the data bus. The use of tri-state buffers in a bus structured system is shown in Figure 4.8.

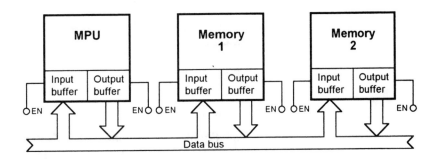

Figure 4.8

Enabling of the tri-state buffers in this system is under the control of the MPU which ensures that at any instant in time:

1   only one output buffer is actively connected to the data bus,
2   only one input buffer is actively connected to the data bus, and
3   input and output buffers of the same device are not simultaneously enabled.

Activity

1   Using breadboard or similar techniques, connect a 74LS244 octal buffer into circuit as shown in Figure 4.9.

2   Use S1 and S2 to apply appropriate logic 0 and logic 1 states to the buffer input and enable pins.

3   Using a logic probe, monitor the state of the output pin for each combination of inputs.

4   Tabulate results as shown in Table 4.2 and record your conclusions.

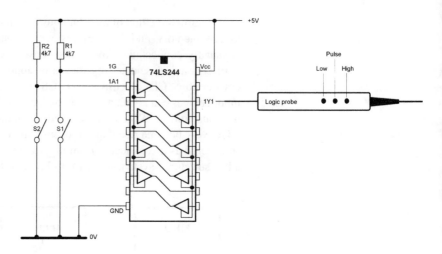

Figure 4.9

Table 4.2

Enable 1G	Input 1A1	Output 1Y1
0	0	
0	1	
1	0	
1	1	

Control bus signals for data transfers

Control signals required for each memory device are:

1 a **chip select ($\overline{\text{CS}}$)** signal to select one particular memory IC,
2 an **$\overline{\text{RD}}$ (read)** signal to enable the output buffer to allow the memory to place information onto the data bus, and
3 a **$\overline{\text{WR}}$ (write)** signal if the memory is RAM to enable the input buffer to allow information from the data bus to be stored in the memory.

Chip select signals are obtained by decoding the high order address lines (*those not directly connected to the memory device and which are otherwise unused*). The $\overline{\text{RD}}$ and $\overline{\text{WR}}$ signals are supplied by the MPU and the $\overline{\text{RD}}$ signal becomes active (low) when the MPU wishes to read data from the memory, while the $\overline{\text{WR}}$ signal becomes active (low) when the MPU wishes to store data in the memory. The MPU ensures that these signals are generated at the correct time. When using ROM, the $\overline{\text{WR}}$ signal is not required because ROMs are read-only memory and have only output buffers.

The connections between an MPU and its memory devices are shown in Figure 4.10.

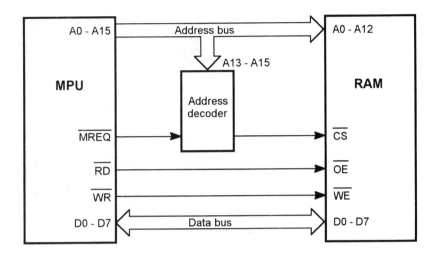

Figure 4.10

Bus buffering

The outputs of a typical microprocessor are only capable of driving a **single TTL load**. Each memory device that is connected to the microcomputer buses represents a **capacitive load** on each line of the bus. This is particularly so with MOS devices, since these have relatively high input and output capacitances (typically 5 pF input capacitance and 10 pF output capacitance). The effect of excessive capacitance on the buses is to slow down the rate at which data can change and thus prevent correct timing of operations. Where many memory and I/O devices are connected to the buses, it becomes necessary to connect a bus buffer between the MPU and its memory and I/O circuits. A tri-state buffer is used, and its function is to supply large drive currents to the buses, yet impose very little loading on the MPU. Such a buffer is often called a **bus driver**. A unidirectional buffer may be used for the address bus, but a bidirectional bus driver is needed for the data bus to allow data transfers to take place between an MPU and its memory or I/O devices in either direction. A typical circuit arrangement is shown in Figure 4.11.

Address decoding   Microcomputer memory chips make use of binary addresses since this results in a reduction in the number of address select lines required for a given memory size. If 'n' address lines are available on a memory chip, then 2^n different binary inputs are possible; therefore 'n' address lines permits 2^n

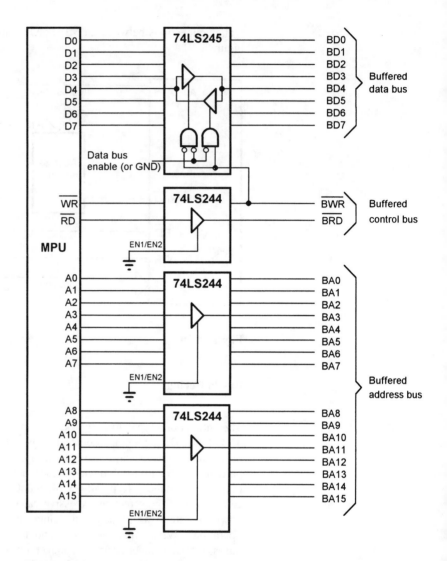

Figure 4.11

different locations to be addressed. For example, if a microcomputer memory has 12 address lines, then the number of different addressable locations is 2^{12} or 4096_{10}. The binary address applied to a memory must first be **decoded** to enable the selection of a single location to take place. Most proprietary memory ICs include some internal address decoding circuits, as shown in Figure 4.12.

Decoding is the use of a logic network to detect unique combinations of binary numbers, each binary input having its own particular output. Consider a two-bit binary number that can be used to represent four possible combinations of '0' and '1'. A particular combination number can be detected by use of the circuit shown in Figure 4.13.

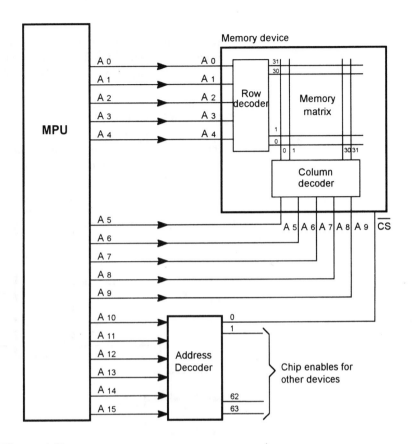

Figure 4.12

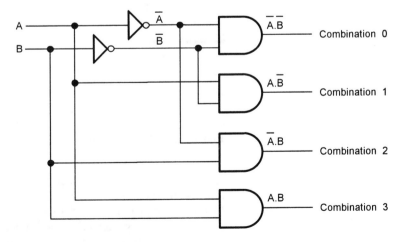

Figure 4.13

It can be seen that the decimal equivalent of the two-bit code determines which combination number is detected. For example, if **B=1** and **A=0** which is equivalent to **decimal 2**, then combination **number 2** is detected because it is at logic state '1' while all other combination numbers are at logic state '0'.

If it is required to detect a particular combination of binary numbers by the use of logic state '0' rather than a logic state '1', then the AND gates in the previous circuit could be replaced by NAND gates. Decoding circuits of this type are known as **n line to 2^n line decoders**, and their operation may be studied by referring to Figure 4.14.

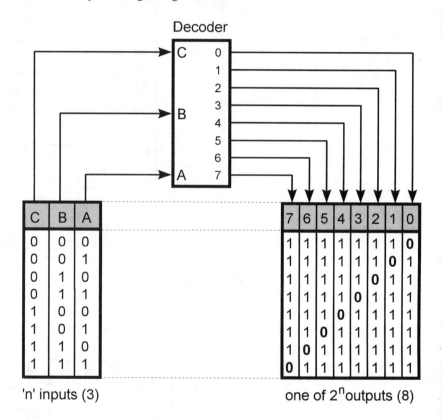

Figure 4.14

Practical address decoding

External decoding in a microcomputer may be carried out using TTL family devices, for example the 74LS139, for which the pin out detail and logic function is shown in Figure 4.15.

A two-bit binary number can be applied to the device using pins A and B, where A is the least significant bit. Output pins Y0, Y1, Y2 and Y3 indicate which binary number has been applied to inputs A and B by a change in logic

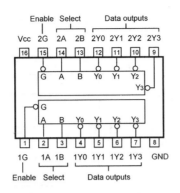

Figure 4.15

74LS139

Inputs			Outputs			
Enable	Select					
G	B	A	Y0	Y1	Y2	Y3
H	X	X	H	H	H	H
L	L	L	L	H	H	H
L	L	H	H	L	H	H
L	H	L	H	H	L	H
L	H	H	H	H	H	L

74LS139

Figure 4.16

level of the appropriate output. It should be clear that the decoder will only process the inputs to give the required outputs if it has been enabled, otherwise there will be no corresponding change in output. Note that this device is a **dual 2 line to 4 line decoder** chip which means that it has two identical and independent decoders of the type shown in Figure 4.16 in one package.

There are a large number of different decoder chips within the 74 TTL series, and a further example is the 74LS138 which is a 3 line to 8 line decoder chip, as shown in Figure 4.17.

74LS138

Inputs					Outputs							
Enable		Select										
G1	G2	C	B	A	Y0	Y1	Y2	Y3	Y4	Y5	Y6	Y7
X	H	X	X	X	H	H	H	H	H	H	H	H
L	X	X	X	X	H	H	H	H	H	H	H	H
H	L	L	L	L	L	H	H	H	H	H	H	H
H	L	L	L	H	H	L	H	H	H	H	H	H
H	L	L	H	L	H	H	L	H	H	H	H	H
H	L	L	H	H	H	H	H	L	H	H	H	H
H	L	H	L	L	H	H	H	H	L	H	H	H
H	L	H	L	H	H	H	H	H	H	L	H	H
H	L	H	H	L	H	H	H	H	H	H	L	H
H	L	H	H	H	H	H	H	H	H	H	H	L

Figure 4.17

Activity

1 Using breadboard or similar techniques, connect a 74LS139 2 line to 4 line decoder into circuit as shown in Figure 4.18.

2 Use S1–S3 to apply appropriate logic 0 and logic 1 states to A and B inputs and G enable pins.

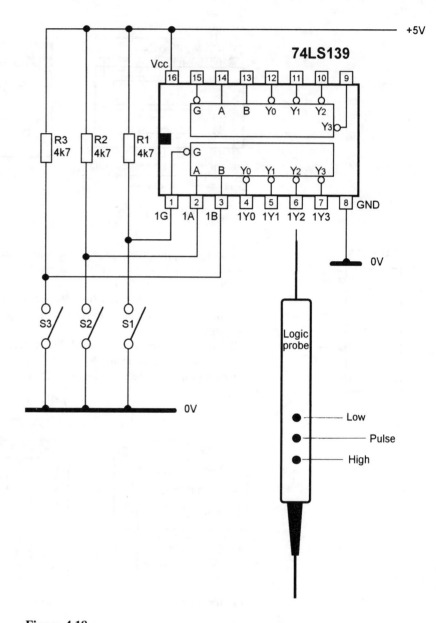

Figure 4.18

3   Using a logic probe, monitor the state of output pins Y0 to Y3 for each combination of inputs.

4   Tabulate results as shown in Table 4.3 and record your conclusions.

Table 4.3

Inputs			Outputs			
Enable	Select					
G	B	A	Y3	Y2	Y1	Y0
1	X	X				
0	0	0				
0	0	1				
0	1	0				
0	1	1				

Memory organization

Practical limitations often mean that it is not possible to use a single memory device for all storage requirements within a microcomputer for the following reasons:

1   different types of memory may be needed, e.g. ROM and RAM,
2   more storage locations than available in a single chip may be needed,
3   insufficient number of bits in a memory chip for the bus width used.

Therefore ways must be found to combine and assign appropriate addresses to devices in a practical memory system.

Increasing number of bits

The number of bits at each location may be increased by connecting memory devices in parallel as in Figure 4.19 which shows how two 8-bit memory devices may be connected to a 16-bit data bus.

   Both devices are simultaneously selected, and each address applied to the address bus selects equivalent locations in both devices which then transfer data to and from their respective bits of the data bus.

Increasing number of locations

The number of memory locations may be increased by connecting memory devices as in Figure 4.20 which shows how two **16K x 8-bit** RAMs may be connected to form a **32K x 8-bit** RAM. The basic address for each device is provided by A_0 to A_{13}, but only one of the devices is enabled at any instant, using chip select signals derived from the decoding of A_{14} and A_{15}.

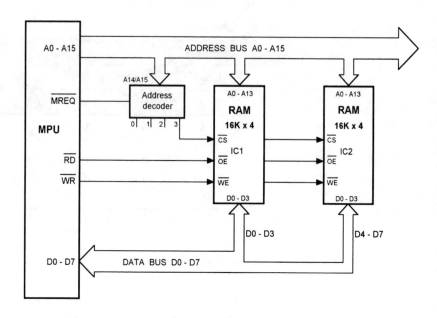

Figure 4.19

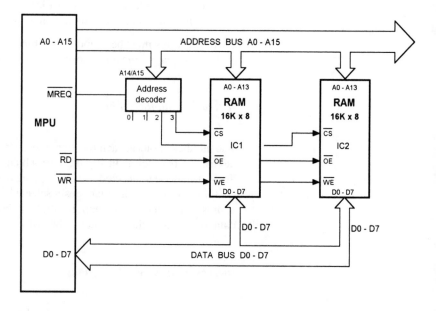

Figure 4.20

Since the decoding of A_{14} and A_{15} provides four select signals, it is possible to extend this technique to give a 64K block of memory.

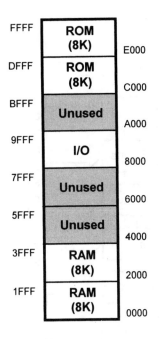

FFFF ROM (8K) E000
DFFF ROM (8K) C000
BFFF Unused A000
9FFF I/O 8000
7FFF Unused 6000
5FFF Unused 4000
3FFF RAM (8K) 2000
1FFF RAM (8K) 0000

Memory map

Figure 4.21

Memory maps

The location of different memory or I/O devices within the available address space depends very much upon the address decoder arrangements and may be represented in graphical form by a **memory map**, as shown in Figure 4.21.

In some systems it is possible for the user to alter the position of devices within the memory map; however, care must be exercised when doing this since a microprocessor expects certain devices to be located in fixed locations. ROM must be located at the address generated by the MPU after reset, for example, and RAM must be located where the system expects to find the stack. System programs called **monitor routines** are sometimes invoked by user's software, and if the addresses of these routines are changed then programs are unlikely to function.

Determining address ranges

Sometimes a memory map or similar documentation is unavailable, and in these cases it may be necessary to determine the address ranges used in a microcomputer. In such cases it is often helpful to consider internal and external decoding, as shown in the following example.

Example

Show how it is possible to uniquely identify any single address within a **16K x 8-bit** memory made up from eight **2K x 8-bit** memory chips, as shown in Figure 4.22.

Internal decoding:

Eleven address lines, A_0 to A_{10}, will be required in order to identify one of the 2048 locations within a memory chip, once it has been selected, i.e. $2^{11} = 2048_{10}$.

External decoding:

Three address lines, A_{11} to A_{13}, will be required in order to select one of the eight memory chips, i.e. $2^3 = 8_{10}$.

Address range:

The address range of each memory device is determined by evaluating its lowest and highest addresses. The lowest address is found by first writing

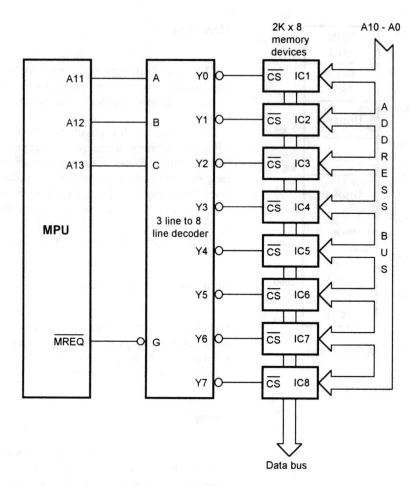

Figure 4.22

down the states of the externally decoded address lines that select the device, followed by logical 0's for each address line that is decoded internally. The highest address is found by repeating this process using logical 1's for the internally decoded address lines. For example, the lowest and highest addresses for IC1 are shown in Figure 4.23.

The actual chip selected is determined by the combination of 0's and 1's set up on the address bus lines A_{11}, A_{12} and A_{13}, provided that the 3 line to 8 line decoder chip has been **enabled**. In some systems the address decoder may be permanently enabled by connecting its G input to ground, but other systems require an enable signal such as $\overline{\text{MREQ}}$ (Z80) or M/$\overline{\text{IO}}$ (8086). It is important to notice that input A of the decoder is connected to the least significant of the address bus lines used for external decoding. Clearly eight combinations of 0's and 1's can be set up on the address lines A_{11}, A_{12} and A_{13}, and in response to each of these combinations, one of the memory chips will be selected by the 3 line to 8 line decoder chip outputs.

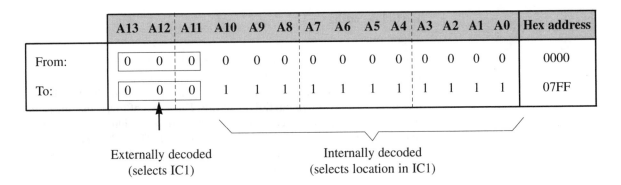

Figure 4.23

The address range for each chip can be determined from Figure 4.24.

	External decoding			Internal decoding											Hex address
	A13	A12	A11	A10	A9	A8	A7	A6	A5	A4	A3	A2	A1	A0	
IC1	0	0	0	0	0	0	0	0	0	0	0	0	0	0	0000
	0	0	0	1	1	1	1	1	1	1	1	1	1	1	07FF
IC2	0	0	1	0	0	0	0	0	0	0	0	0	0	0	0800
	0	0	1	1	1	1	1	1	1	1	1	1	1	1	0FFF
IC3	0	1	0	0	0	0	0	0	0	0	0	0	0	0	1000
	0	1	0	1	1	1	1	1	1	1	1	1	1	1	17FF
IC4	0	1	1	0	0	0	0	0	0	0	0	0	0	0	1800
	0	1	1	1	1	1	1	1	1	1	1	1	1	1	1FFF
IC5	1	0	0	0	0	0	0	0	0	0	0	0	0	0	2000
	1	0	0	1	1	1	1	1	1	1	1	1	1	1	27FF
IC6	1	0	1	0	0	0	0	0	0	0	0	0	0	0	2800
	1	0	1	1	1	1	1	1	1	1	1	1	1	1	2FFF
IC7	1	1	0	0	0	0	0	0	0	0	0	0	0	0	3000
	1	1	0	1	1	1	1	1	1	1	1	1	1	1	37FF
IC8	1	1	1	0	0	0	0	0	0	0	0	0	0	0	3800
	1	1	1	1	1	1	1	1	1	1	1	1	1	1	3FFF

Figure 4.24

Partial decoding

If the circuit shown in Figure 4.22 is used with an MPU having 16 address lines it will be noticed that some of them are unused, i.e. A_{14} and A_{15}. In any microcomputer, when some of the address lines are unused, the decoding is incomplete and is called **partial decoding**. The effect of this is that the same memory device may be accessed at several different addresses, depending upon how many of the address lines are unused. This effect is known as **imaging** or **ghosting** and is best avoided by using full address decoding. This may be achieved by the use of additional cascaded decoding chips as shown in Figure 4.25.

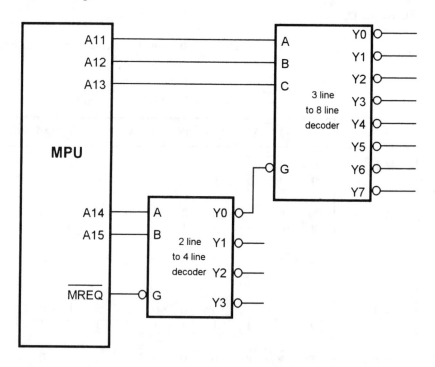

Figure 4.25

Activity

Construct logic tables similar to Figure 4.24 and hence determine the address ranges selected by:

1   Y0 to Y3 of the 2 line to 4 line decoder, and

2   Y0 to Y7 of the 3 line to 8 line decoder.

Practical memory circuit

When working with microcomputer circuits it is sometimes necessary to add extra memory to the system. With desktop computers this can usually be achieved by purchasing extra SIMMs (single in-line memory modules) and slotting them into connectors already provided on the system motherboard. Small microcontroller systems may not offer such facilities, and in such cases it becomes necessary to design small memory expansion circuits which are mapped into the system at any suitable free block in the memory map.

Example

The memory map of a microcomputer system is shown in Figure 4.26(a). Design a 2K x 8-bit memory circuit to provide an expansion of the system RAM using devices shown in Figure 4.26(b) and (c).

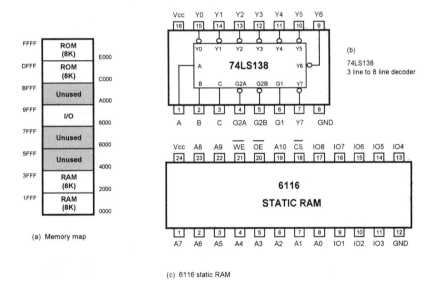

(a) Memory map

(b) 74LS138
3 line to 8 line decoder

(c) 6116 static RAM

Figure 4.26

The 6116 is a **2K x 8-bit static RAM** with inputs A0 to A10 which are decoded internally to select its different locations. This means that the five MPU address lines **A11 to A15** must be decoded externally to provide a chip select signal to map the RAM to addresses **4000H** to **47FFH**. The 74LS138 device is a 3 line to 8 line decoder which implies that it can only decode *three* address lines. However, there are three enable inputs on this decoder, G1, $\overline{G2A}$ and $\overline{G2B}$, therefore it is possible to use two of these inputs in the decoding process so that full address decoding is achieved, as shown in Figure 4.27.

External decode 74LS138					Internal decode 6116											Hex address
A15	A14	A13	A12	A11	A10	A9	A8	A7	A6	A5	A4	A3	A2	A1	A0	
0	1	0	0	0	0	0	0	0	0	0	0	0	0	0	0	4000H
0	1	0	0	0	1	1	1	1	1	1	1	1	1	1	1	47FFH
$\overline{\text{G2A}}$	G1	C	B	A												
Fixed		Use Y0 output														

Figure 4.27

In some systems it is necessary to include a memory select signal such as $\overline{\text{MREQ}}$ (Z80) or M/$\overline{\text{IO}}$ (8086) to enable the decoder circuit, and if needed, one of the G inputs must be reserved for this purpose.

The control bus signals $\overline{\text{RD}}$ and $\overline{\text{WR}}$ are connected to the 6116 $\overline{\text{OE}}$ and $\overline{\text{WE}}$ inputs respectively to enable its output and input buffers during read and write cycles. A circuit of the final design is shown in Figure 4.28.

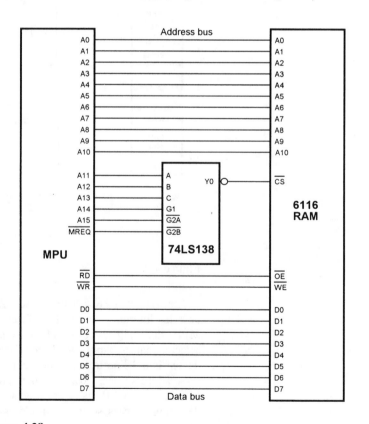

Figure 4.28

Activity

If you have available a small microprocessor system with unallocated space in the memory map, and access to the system buses:

1   Design a 2K x 8-bit memory expansion circuit for your system using a 6116 RAM and 74LS138 decoder.

2   Build your circuit using breadboard or similar construction.

3   Connect your circuit to the microcomputer system buses.

4   Test your circuit by storing and retrieving data from selected locations and checking that no conflicts exist.

5   Write a report to include your logic table, circuit diagram, test instructions and data, and notes concerning any problems encountered.

Test your knowledge 4.2

1   A bus conflict occurs when two devices simultaneously try to drive a data line to:
   A   logical 0
   B   logical 1
   C   the same logic state
   D   opposite logic states

2   When disabled, the output state of a tri-state buffer becomes:
   A   logical 0
   B   high impedance
   C   logical 1
   D   low impedance

3   In a bus structured system, tri-state buffers are:
   A   selectively enabled to prevent bus conflicts
   B   all disabled to allow bus conflicts
   C   all enabled to prevent bus conflicts
   D   selectively enabled to allow bus conflicts

4 An active $\overline{WR}$ signal allows:
   A   a RAM to write to the data bus
   B   a ROM to write to the data bus
   C   an MPU to write to the address bus
   D   an MPU to write to the data bus

5 An active $\overline{RD}$ signal enables:
   A   an MPU to read the data bus
   B   a RAM to read the data bus
   C   an MPU to read the address bus
   D   a ROM to read the data bus

6 The loading of microcomputer buses is minimized by using:
   A   unidirectional address and unidirectional data bus buffers
   B   bidirectional address and bidirectional data bus buffers
   C   bidirectional address and unidirectional data bus buffers
   D   unidirectional address and bidirectional data bus buffers

7 When connecting a 2K x 8-bit memory device to the buses of a Z80 microcomputer, the number of address lines to be decoded externally is:
   A   16
   B   8
   C   11
   D   5

8 When a 3 line to 8 line decoder is used to decode the highest three address lines of a 16-bit address bus, select signals are generated for:
   A   eight 8K blocks of memory
   B   three 8K blocks of memory
   C   eight 16K blocks of memory
   D   three 16K blocks of memory

9 Each memory location in a microcomputer responds to four different addresses, therefore address decoding is:
   A   partial and two address lines are left unused
   B   full and all address lines are used
   C   partial and four address lines are left unused
   D   full and two address lines are left unused

10 Two 4K x 8-bit RAMs are connected in parallel except for their CS pins, which are connected to different outputs of an address decoder. The resulting RAM size is:

A 4K x 8-bit
B 4K x 16-bit
C 8K x 8-bit
D 8K x 16-bit

5 Software development

Summary

This element deals with writing software to solve simple engineering problems using structured design techniques. The use of Jackson Structured Programming (JSP) is covered at an elementary level and shows how programs are written using machine code or assembly language, although a structured design approach does not limit the user to programming at this level. Finally, software testing procedures are introduced, showing how programs may be checked for correct operation.

Program development

Software is the term used to describe programs which may be loaded into a microcomputer RAM to give the system its designated operating characteristics. Ideally, software should be developed so that it operates reliably at all times, functioning as intended, and be written so that it is easily maintained. These requirements are unlikely to be achieved if program development is based upon inspirational or trial and error types of techniques. In this respect, software development is no different from hardware development. It therefore follows that some form of *discipline*, or *engineering principles*, must be applied to software development, i.e. **software engineering**. This is particularly true when developing large and complex programs involving a number of different programmers.

Program design sequence

Although the programs dealt with in this element are relatively small, as a matter of good practice, correct design techniques should be used and the temptation to design and enter programs 'on-screen' simultaneously should be avoided. A sequence that is commonly adopted consists of the following steps:

1 **Problem definition** (program requirements)
2 **Algorithm**
3 **Flowchart or structure chart**
4 **Coding**
5 **Test and debug**

Problem definition

The nature of a problem to which a solution is being sought must be clearly defined before programming can take place, otherwise the resultant program may not perform its required task adequately. An eloquent program that solves the wrong problem is of no use at all. A program requirement specification may be prepared at this stage to include the following points:

1 **state** precisely what **functions** the program is to perform,
2 **list** exact number of **inputs** and **outputs** required,
3 **state** any **constraints** regarding speed of operation, accuracy, memory requirements etc.,
4 **action** to take **if errors occur**.

Algorithm

An algorithm consists of a sequence of actions which define a method of solving a particular problem. It is possible for a problem to have more than one algorithm, i.e. more than one way of solving it. The lack of a suitable algorithm indicates that a problem may have no solution.

Flowchart or structure chart

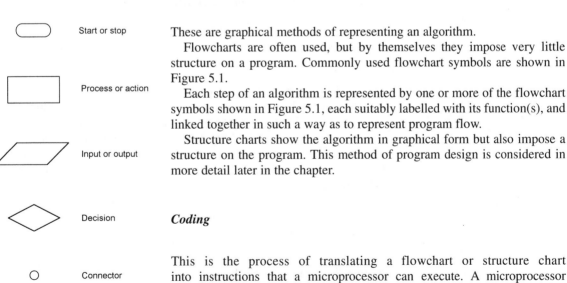

Start or stop

Process or action

Input or output

Decision

Connector

Flow direction

Figure 5.1

These are graphical methods of representing an algorithm.

Flowcharts are often used, but by themselves they impose very little structure on a program. Commonly used flowchart symbols are shown in Figure 5.1.

Each step of an algorithm is represented by one or more of the flowchart symbols shown in Figure 5.1, each suitably labelled with its function(s), and linked together in such a way as to represent program flow.

Structure charts show the algorithm in graphical form but also impose a structure on the program. This method of program design is considered in more detail later in the chapter.

Coding

This is the process of translating a flowchart or structure chart into instructions that a microprocessor can execute. A microprocessor is only capable of interpreting binary instructions known as **machine code**. This form of coding is rather tedious and error-prone for human operators, therefore machine code programs are usually written either in hexadecimal codes, or by using symbolic notation known as **assembly language**.

Testing and debugging

Testing of a program may take place at various levels, often using paper and pencil techniques such as **desk check**, **dry run** or **trace tables**. However, once the program code has been written, it must eventually be loaded into memory and tested for correct operation. In order to load a simple machine code program into memory, some software must already exist within the machine, and this is usually called a **monitor program** or a **debugger**. Except for very simple routines, few programs perform as intended at the first attempt. Various errors may occur in the program writing processes that lead to partial or total malfunctioning of a program. These errors are generally known as **bugs** and the process of tracking down and eliminating bugs is known as **debugging**.

Structured programming

Structured programming is a software design method, consisting of a basic set of principles and techniques, that enables problems to be solved by a computer. The method used should enable correct programs to be produced, large or small, so that each stage of development is within the intellectual limits of the programmer. Several different types of structured programming exist, but all program design methods are based upon two fundamental principles:

1 **stepwise refinement**, and
2 **three basic program control constructs**.

Stepwise refinement

The object of program design is to take the program requirement specification and break this down into smaller items suitable for translation into instructions for the chosen programming language. This process usually takes place step by step, **breaking down** (or *refining*) the main function into smaller functions (subfunctions) which are each further broken down until the lowest level is reached. This is shown in Figure 5.2 and is generally called **top-down design**.

Program constructs

Program constructs are the basic building blocks of programs. There are only three basic program constructs and these are **sequence**, **iteration** and **select**, each of which should have only one entry point and one exit. Flowchart arrangements for each construct are shown, and if used, these will impose some form of structure on a program flowchart.

These basic constructs are described in the following subsections.

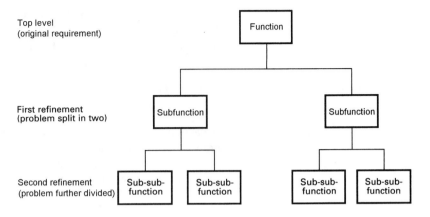

Figure 5.2

Sequence (linear)

A sequence is a program function consisting of two or more parts (subfunctions) and program flow is linear, from one part to the next (i.e. there is no choice or repetition), as shown in Figure 5.3.

An entire program may be just a sequence of subfunctions although each subfunction may contain other constructs. It is also conceivable that very simple machine code programs may use this construct only.

Iteration (loop)

An iteration is a program function with one subfunction that is executed zero or more times, i.e. a **program loop**. The number of times that the subfunction is executed depends upon a condition imposed upon a control variable within the looping process. For example, consider the example shown in Figure 5.4.

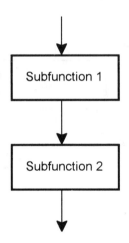

Figure 5.3

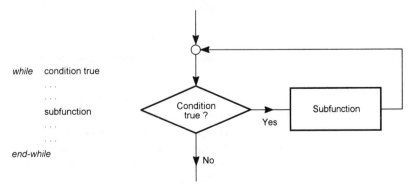

Figure 5.4

In this example, if the variable is a non-zero value then the subfunction will be repeated continuously. Once the variable becomes zero, the subfunction will be completed but there will be no further repetition.

Selection (choice)

A selection construct is a program function that has two or more subfunctions, one of which is executed once only. Selection of a particular subfunction is based upon a specified condition being true. For example, consider the example shown in Figure 5.5.

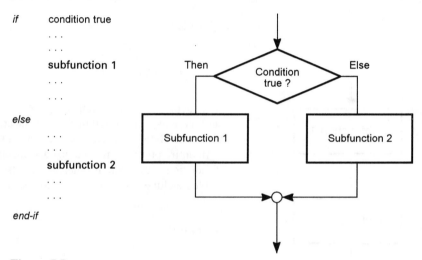

```
if      condition true
        . . .
        . . .
        subfunction 1
        . . .
        . . .

else
        . . .
        . . .
        subfunction 2
        . . .
        . . .

end-if
```

Figure 5.5

Either **subfunction 1** or **subfunction 2** will be executed once only, depending whether the condition is true or not. It is possible for this construct to have a large number of different choices, and the 'else' condition is that which does not satisfy any of these other choices.

Jackson Structured Programming (JSP)

There are a number of different structured program design methods in use. One of these methods, **JSP** or **Jackson Structured Programming**, was designed by Michael Jackson and is based upon the two fundamental principles of structured programming already outlined, i.e. stepwise refinement and three basic construct types.

Program design is represented by a hierarchical structure diagram in JSP, as shown in Figure 5.6.

Top level
(original requirement)

Binary to
BCD
program

First refinement

①

Convert
body

⑤

Second refinement

Input binary
number

Convert
to BCD

Display

Third refinement

②

③

④

1, 2, 3, 4 and 5 are the lowest levels of refinement for each subfunction

Figure 5.6

Each level of the structure diagram represents a different level of refinement, and the following notation is used in JSP for showing the refinement of each construct type.

Select

The JSP notation for a sequence is shown in Figure 5.7.

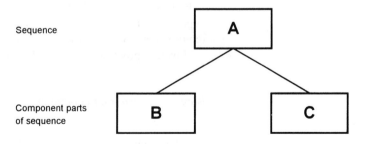

A is a sequence of two subfunctions, B and C

Figure 5.7

Function A is a **sequence** consisting of **subfunction B** followed by **subfunction C**. If subfunctions B and C have no further definition, they are known as **elementary program component**s, otherwise they contain other program constructs which should be further refined.

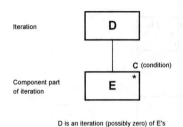

Figure 5.8

Iteration (loop)

The JSP notation for iteration is shown in Figure 5.8.

Function D is an **iteration** of **subfunction E** while **condition C is true**, i.e. D consists of repeated execution of E while condition C remains true. Note the use of the **asterisk symbol** (*) in box E to indicate that box D is an iteration, together with identification of the controlling condition. Subfunction D may be an elementary program component or may contain further program constructs.

Selection (choice)

The JSP notation for selection is shown in Figure 5.9.

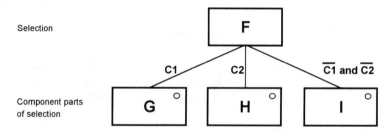

Figure 5.9

Function F is a **selection** of one of the **subfunctions G, H** and **I**, depending upon conditions C1 and C2 which are defined according to the requirements of the program. Note the use of the circle (o) in each subfunction box to indicate that F is a selection.

Elementary program components

An elementary component is not a control function but must be used in conjunction with the three control components already described to construct a program. An elementary component requires **no further refinement** and represents a sequence of actions that can be readily achieved in the chosen programming language. These actions are indicated by numbered boxes or circles, as in Figure 5.10 which shows J as a sequence of actions, 1 followed by 2 followed by 3.

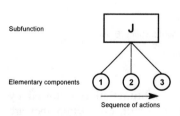

Figure 5.10

Incorrect constructs

At certain stages of refinement, particularly when allocating actions to a program structure diagram, **invalid constructs** of the type shown in Figure 5.11 may result.

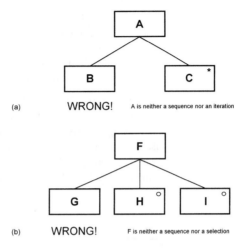

(a)   WRONG!   A is neither a sequence nor an iteration

(b)   WRONG!   F is neither a sequence nor a selection

Figure 5.11

These are incorrect constructs since a function cannot be a mixture of different types of constructs. This problem may be dealt with by the inclusion of an additional block, called a function **body**, the purpose of which is to maintain the correct structure of the JSP chart. The incorrect constructs of Figure 5.11 may be modified as shown in Figure 5.12.

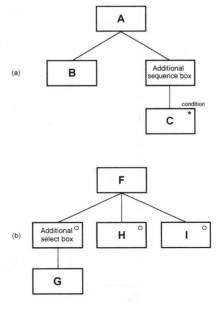

Figure 5.12

JSP design method

The JSP design method consists of a number of defined steps, and these are:

Step 1   Produce data structure diagrams
Step 2   Produce preliminary version of program structure diagram
Step 3   Identify and list conditions and actions
Step 4   Allocate actions and conditions to program structure diagram
Step 5   Produce final version of program structure diagram

Simple engineering problems of the type covered in this chapter will have very elementary data structures, typically parallel data inputs, some form of processing, followed by parallel data outputs. In such cases the data and preliminary program structure diagrams are likely to be identical, therefore Step 1 is likely to be omitted, or combined with Step 2 . A typical design sequence is illustrated by the following example.

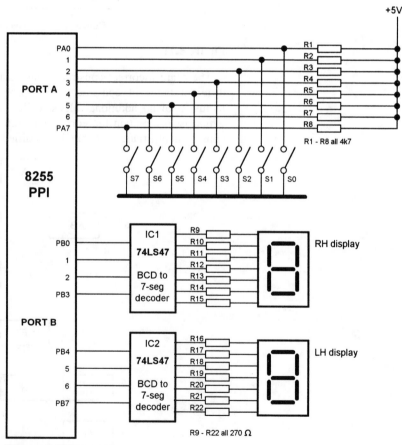

Figure 5.13

Worked example

A microcomputer system has an 8255 PPI interface with 8-bit parallel I/O Ports A and B connected to switches and 7-segment displays, as shown in Figure 5.13.

A program is required to carry out the following operations:

1   read a binary input from the switches,
2   convert binary to BCD by adding 6 if a number greater than 1001_2 (9_{10}) is entered,
3   display the BCD result on the 7-segment displays,
4   terminate the program if a binary input greater than 1111_2 (15_{10}) is entered.

Step1/Step2 – Preliminary program structure diagram

A preliminary program structure diagram is shown in Figure 5.14.

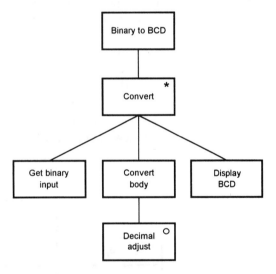

Figure 5.14

The preliminary program structure diagram shows the program as a conditional iteration of the sequence, **get binary input – convert – display BCD**, with a conditional selection of **decimal adjust**.

Step 3 – Identify and list conditions and actions

A list of conditions and actions must be prepared at this stage, and these may be later cross-referenced with the structure diagram. When defining conditions for an iteration, the following questions should be answered:

1   What condition allows iteration to continue?
2   What condition terminates iteration?

When defining conditions for a selection, the following questions should be answered:

1 What must be true to select a particular option?
2 Does the condition in (1) select just one option?
3 What happens if no conditions are true?

This program is relatively simple and needs just one condition for the iteration, and one for the selection, identified as C1 and C2.

C1 is a condition that allows iteration while the input binary number is in the range 0000_2 to 1111_2 (15_{10}). Iteration ceases and the program is terminated if a number greater than 1111_2 is entered.

C2 is a condition that adds 0110_2 (6_{10}) to the input binary number if it is greater than 1001_2 (9_{10}), otherwise the number is left unchanged.

Therefore the conditions list is as follows:

(a) C1 While input $< = 1111_2$ (15_{10})
(b) C2 If input $> 1001_2$ (9_{10})

Elementary program actions can be identified by reference to the original program specification, and may be dealt with in the following order:

(c) Initialization and termination of the program
(d) Inputs
(e) Outputs
(f) Processing of data, input to output
(g) Operations to support the conditions list

Applying these techniques to this program gives the conditions and action list shown in Figure 5.15.

CONDITIONS	
C1 While input $< = 1111$	
C2 If input > 1001	
STEP	**ACTIONS**
Initialisation and termination	1. Configure the 8255 PPI - Port A input, Port B output
	2. Return to the operating system (system dependent)
Input	3. Read Port A switches
Output	4. Write BCD to the 7-segment displays (Port B)
Processing	5. Add 06 to the switch reading
Operations to support conditions	6. Check if binary input is greater than 1111
	7. Check if binary input is greater than 1001

Figure 5.15

Step 4 – Allocate conditions and actions to program structure diagram

Allocating conditions and actions identified in Figure 5.15 results in the structure diagram shown in Figure 5.16.

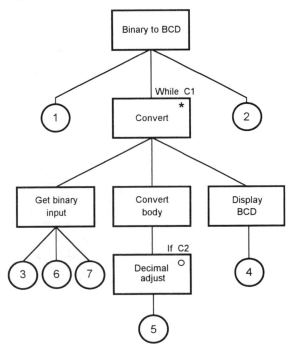

Figure 5.16

It may be noticed that the addition of actions has disturbed the structure of this diagram, since the block marked **Binary to BCD** now appears to be both a sequence and an iteration. This incorrect construct must now be corrected.

Step 5 – Produce final version of program structure diagram

The program structure shown in Figure 5.16 may be corrected by adding a box marked **Binary to BCD body**, as shown in Figure 5.17.

The effect of adding this box is to restore the box marked **Binary to BCD** to the status of a sequence.

Coding

Although it is possible to convert a JSP structure diagram directly into program code, it is advisable to first make use of schematic logic which is a **pseudo-code** (similar to high-level language) representation of the program

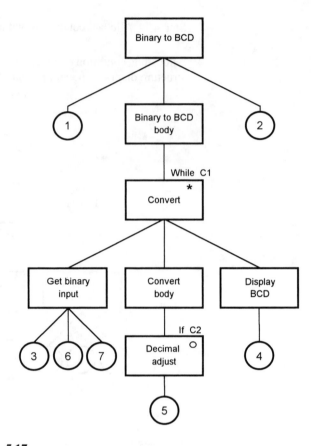

Figure 5.17

structure. Schematic logic for each of the basic constructs may be illustrated by using our binary to BCD program as an example.

(a) **Sequence**

```
BINARY TO BCD          seq     {construct_name, type}
      DO    1                  {configure 8255 PPI}
      . . .                    {Binary to BCD body}

      . . .
      DO    2                  {return to operating system}
BINARY TO BCD          end
```

(b) **Iteration**

```
BINARY TO BCD BODY  iter  WHILE C1   {construct_name, type,
                                       condition}
        CONVERT    seq
```

```
                DO   3        {read switches}
                . . .
                DO   4        {write BCD to 7-segment displays}
        CONVERT   end
BINARY TO BCD BODY  end
```

(c) **Selection**

```
CONVERT BODY   sel      IF C2 {construct _name, type,
                                condition}
        DO   5                  {add 06 to input number}
CONVERT BODY   end
```

The schematic logic for the complete program may now be formed by completing these constructs and combining as necessary. Note that all conditions and actions should be present in the final schematic logic, and this may readily be checked.

```
BINARY TO BCD  seq
    DO   1                  {configure 8255 PPI}
    BINARY TO BCD BODY   iter    WHILE C1 {input
                                  number <= 15}
        CONVERT   seq
            DO   3              {read switches}
            DO   6              {check for number > 15}
            DO   7              {check for number > 9}
            CONVERT BODY   sel IF C2 {if number > 9}
                DO   5          {add 06 to input number}
            CONVERT BODY   end
            DO   4              {write BCD to 7-seg
                                displays}
        CONVERT   end
    BINARY TO BCD BODY   end
        DO   2              {return to operating system}
BINARY TO BCD   end
```

At this stage the program is ready for coding into the programming language being used (the target language), which could be a high-level language such as **Pascal** or **C**; however, this element deals with programs at **machine code** or **assembly language** level.

Converting the JSP structure chart into program code is a relatively simple matter, and code for the Z80, 8086 and 8031 microprocessors is shown in Figure 5.18.

The user RAM and Port I/O addresses relate to equipment available to the author, but may be readily changed for alternative equipment. Note how each subfunction in the original program structure converts into one or two machine code instructions.

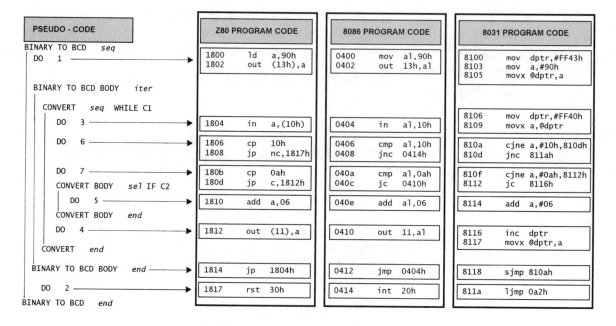

Figure 5.18

The actual machine code for this program will be required for testing purposes and this may be obtained by either:

1 looking up equivalent hex codes in the instruction set, or
2 using an assembler to automatically generate the codes.

Method 2 is preferable since it is quicker and less error-prone. Many microprocessor monitor or debugging programs include an **in-line assembler** which allows code to be entered as shown in Figure 5.18. A better option, however, is to use a **full assembler** since this allows the use of **symbolic names** to replace hex numbers, **labels** to identify addresses, and **comments** to assist in program documentation. Programs are prepared using a **text editor** and are saved as text files which may then be processed by the assembler. Errors may be corrected by editing a text file followed by re-assembly.

Assembly language

Assembly languages are unique to their particular MPU, although apart from the use of different mnemonics, many assemblers for different types of MPU often have much in common. On the other hand, there appears to be little standardization, and many minor differences will be encountered between different assemblers for the same MPU. Therefore frequent reference to the assembler handbook may be necessary initially.

Assembly language programs consist of a **sequence of statements** (or lines), with each statement divided into a maximum of **four fields**. A field consists of a **group of characters**, and fields are separated from each other by at least one space or tab character, except for the comment field which is usually preceded by a **semicolon** (;). The fields of a typical assembly language program are as follows:

LABEL FIELD	OPERATOR FIELD	OPERAND FIELD	;COMMENT FIELD
e.g. start:	mov	cx,0ec58h	;initialize delay

Labels

A label is a sequence of characters which act as an **identifier**, and represent an **instruction address** (or offset) that may need to be referenced by other instructions within a program. Labels are used in place of hexadecimal addresses for two reasons:

1   at the time of writing the source code, the actual offset may not be known, and
2   the use of meaningful labels is an aid to following the flow of a program.

As an example of the use of a label, consider the following 8086 program fragment:

delay:	dec	bl	; reduce count by 1
	jnz	delay	; repeat until zero

At the time of writing the source code, the offset address of **dec bl** is unknown but is assigned the label **delay**. During assembly, the relative offset of **delay** may be determined and the appropriate numerical operand assigned to the **jnz** instruction.

Labels may often be any length that the programmer chooses; however, an assembler usually only accepts a fixed number of characters, and any number in excess of this will be ignored. Labels may use **letters, digits (0–9)** and **special characters such as ? @ _ $**, but the first character of any label must not be a digit (this enables an assembler to distinguish between label names and numbers).

Operators

The operator field may contain a valid **mnemonic** (*instruction*) or an **assembler directive** (*pseudo-op*). There are a large number of different assembler directives, each of which are instructions for the assembler rather than for the MPU, and they are used to set symbolic names to specified values during assembly, specify starting addresses in the program, define storage areas, etc. Assembler directives vary considerably from one assembler to another, but some of the more common directives are described in the following paragraphs.

(i) **.MODEL** (8086 only)

Describes the **memory model** that is to be used (TINY, SMALL, COMPACT, MEDIUM, LARGE, HUGE or FLAT) and is usually the first statement in a program. This defines the size and organization of code and data segments in a program. For a small system the following statement may be used:

.MODEL	**small**	**; small memory model**

(ii) **.STACK** (8086 only)

This directive creates a **stack segment**. By default the assembler allocates 1K of memory for the stack which is sufficient for most programs. Other sizes of stack may be allocated by the following statement:

.STACK	**2048**	**; use 2K stack**

(iii) **.DATA** or **DSEG**

This directive instructs the assembler to create a **data segment** which is used to store the most frequently used program data and keeps it separate from the program codes.

(iv) **.CODE** or **CSEG**

This directive instructs the assembler to start a **code segment** which is used to store the program code. This directive enables all program code to be kept together and prevents it from being mixed with data. The next segment directive closes the previous segment.

(v) **.END** or **END**

This directive is optional with some assemblers and mandatory with others, but if used it must appear as the last statement of a program to inform the assembler that there is **no more code**. It also serves to close any remaining segments where appropriate. Depending upon the way in which a program is written, the END directive may also need an operand, as shown in the following 8086 example:

```
                    .CODE           ; start of code segment
start:              *               ; first instruction here
                    *
                    *
                    *
                    END             start
```

(vi) .ORG or ORG

This directive is used to **initialize the address** of the first instruction that follows. For example, the following program fragment shows initialization of an 8086 code segment offset address to 0400h:

```
                    .CODE           ; start of code segment
                    ORG     0400h   ; start at offset 0400h
start:              *               ; first instruction here
                    *
```

(vii) .EQU, EQU or =

This is a **data assignment** directive that is used to assign a numeric constant to a symbol during assembly (not during execution). The following 8086 example shows how this directive may be used:

```
number1     EQU         5           ; constant 5
number2     EQU         3           ; constant 3
            .DATA
            .CODE
            ORG     0400h
start:      *
            *
            mov         al,number1  ; al <— 05
            add         al,number2  ; al <— al + 03
```

(viii) DB or DEFB

This directive is used to allocate **memory space** for an **8-bit unsigned integer** in the range 0 to 255. For example, the following statement allocates memory space for a variable called small_number, and initializes it to 5:

```
                    .DATA
small_number        DB      5       ; initialize byte to 5
                    .CODE
```

(ix) **DW** or **DEFW**

This directive performs the same function as the DB directive except that it allocates space for a **16-bit unsigned integer**, e.g.

	.DATA		
big_number	DW	2048	; initialize word to 2048
	.CODE		

Operands

Most instructions require one or more operands which define the source or destination of data for use by the instruction. This may be a register, constant or memory address. Constants and memory addresses may be expressed as **numerical constants**, **labels** or **expressions**.

(i) **Numerical constants**

All numeric values are **decimal** by default. This may be changed by either:

    (a) having the directive .**RADIX** *n* or **RAD** *n* at the start of the program, where *n* is 2, 8 or 16.

    (b) appending a **radix indicator** to the numeric value, for example:

10111001b	or	10111001B	binary
25d	or	25D	denary
3fh	or	3FH	hexadecimal

(ii) **Labels, symbols and variables**

One of the major advantages of programming in assembly language rather than machine code is that **labels, symbols** and **variables** may be used rather than hex codes, which has the effect of making a program more readable. They may be defined as follows:

1   a LABEL represents the address of CODE,
2   a SYMBOL represents a CONSTANT,
3   a VARIABLE represents the address of DATA.

The use of these as operands is shown in the following 8086 program fragment:

symbol	EQU	5	;'symbol' represents constant 5
	.CODE		
label:	mov	al,symbol	;'label' represents address
			; of code 'mov al,symbol'

```
            mov     [variable],al  ;'variable' represents data at
                                   ; address represented by 'variable'
            jmp     label          ; jump to address represented by
                                   ; 'label'
variable  DB      0               ; variable initialized to zero
```

These are evaluated at assembly time and their numeric equivalents are substituted.

(iii) **Expressions**

The operands described may be combined in normal algebraic notation using any combination of properly formed operands, operators and expressions, as shown in the following 8086 example:

```
number1    EQU           5            ; number1 = 5
number2    EQU           3            ; number2 = 3
           .CODE
start:     *                          ; first instr
           *
           mov    al,number1+5        ; al <— 10
           mov    al,number1*2        ; al <— 10
           mov    al,number1+number2  ; al <— 8
           mov    al,number1 AND 3    ; al <— 1
```

Reserved words

A reserved word has a special meaning fixed by the language which means that it may only be used under certain conditions. Therefore such words must not be used as labels or symbols in a program. In general this includes all instruction mnemonics and assembler directives; for example, ADD or END should not be used as labels.

Comments

Comments are used for **documentation** purposes only. They are usually preceded by a **semicolon (;)** and are ignored by the assembler. Comments are optional, but their use is essential for good program documentation. Whole line comments may be used to break up a program into its logical stages. Where very long comments are required, an assembler directive such as .COMMENT may be available to avoid repeated use of the semicolon.

Assembly language listings

The following listings for this program were created using full assemblers for each MPU.

```
; ------------------------------------------------
; Z80 Binary to BCD program
; PROGRAM 1
; Program reads an 8-bit binary input, converts
; any number in the range 0000 to 1111 into
; BCD and writes the result to a two-digit
; 7-seg display.
; Entering any number greater than binary
; 1111 terminates the program
; ------------------------------------------------

              ; define all constants used in the program

0010 =            ppi          EQU       10h          ; base address of 8255 PPI
0010 =            switches     EQU       ppi          ; Port A = switches
0011 =            display_led  EQU       ppi+1        ; Port B = 7-seg displays
0013 =            control_reg  EQU       ppi+3        ; 8255 PPI control register
0090 =            io_defn      EQU       90h          ; I/O mode select byte
0010 =            too_big      EQU       16           ; maximum valid number +1
000A =            ten          EQU       10           ; first hex number above 9
0006 =            plus_six     EQU       6            ; decimal adjust factor

1800                           ORG       1800h        ; start of user RAM

              ; configure 8255 PPI, input Port A, output Port B

1800 3E 90       start:        ld        a,io_defn    ; select operating mode
1802 D3 13                     out       (control_reg),a ; and tell PPI

              ; read binary input switches

1804 DB 10       convert:      in        a,(switches) ; get input number

              ; check if binary number greater than 1111, i.e. exit?

1806 FE 10                     cp        too_big      ; finished ?
1808 D2 17 18                  jp        nc,exit      ; if so then quit

              ; check if number greater than 9, if so, decimal adjust

180B FE 0A                     cp        ten          ; need correcting?
180D DA 12 18                  jp        c,below_ten

              ; decimal adjust any number in the range 1010 to 1111
              ; and display the result

1810 C6 06                     add       a,plus_six   ; correction
1812 D3 11       below_ten:    out       (display_led),a ; show result

              ; repeat while input not greater than 1111
              ; otherwise exit program

1814 C3 04 18                  jp        convert      ; iterate while < 16
1817 F7          exit:         rst       30h          ; return to monitor prog

                               END       start
```

```
                              ; --------------------------------------------
                              ; 8086 Binary to BCD program
                              ; PROGRAM 2
                              ; Program reads an 8-bit binary input, converts
                              ; any number in the range 0000 to 1111 into
                              ; BCD and writes the result to a two-digit
                              ; 7-seg display.
                              ; Entering number > binary 1111 terminates
                              ; the program
                              ; --------------------------------------------

                              .MODEL      SMALL          ; code/data segs = 64K
                              .STACK                     ; default stack 1024 bytes

                              ; define all constants used in the program

=     0010     ppi           EQU         10h            ; base address of 8255 PPI
=              switches      EQU         ppi            ; Port A = switches
=     0011     display_led   EQU         ppi+1          ; Port B = 7-seg displays
=     0013     control_reg   EQU         ppi+3          ; 8255 PPI control register
=     0090     io_defn       EQU         90h            ; I/O mode select byte
=     0010     too_big       EQU         16             ; maximum valid number +1
=     000A     ten           EQU         10             ; first hex number above 9
=     0006     plus_six      EQU         6              ; decimal adjust factor

                              .CODE                      ; start of code segment
0400                          ORG         0400h          ; start of user RAM

                              ; configure 8255 PPI, input Port A, output Port B

0400  B0 90    start:        mov         al,io_defn     ; select operating mode
0402  E6 13                  out         (control_reg),al; and tell PPI

                              ; read binary input switches

0404  E4 10    convert:      in          al,(switches)  ; get input number

                              ; check if number greater than 1111, i.e. exit?

0406  3C 10                  cmp         al,too_big     ; finished ?
0408  73 0A                  jnc         exit           ; if so then quit

                              ; check if number > 9, if so, decimal adjust

040A  3C 0A                  cmp         al,ten         ; need correcting?
040C  72 02                  jc          below_ten

                              ; decimal adjust number in range 1010 to 1111
                              ; and display the result

040E  04 06                  add         al,plus_six    ; correction
0410  E6 11    below_ten:    out         (display_led),al; show result

                              ; repeat while input not greater than 1111
                              ; otherwise exit program

0412  EB F0                  jmp         convert        ; iterate while < 16
0414  CD 20    exit:         int         20h            ; return to monitor prog

                              END         start
```

```
                            ; ------------------------------------------------
                            ; 8031 Binary to BCD program
                            ; PROGRAM 3
                            ; Program reads an 8-bit binary input, converts
                            ; any number in the range 0000 to 1111 into
                            ; BCD and writes the result to a two-digit
                            ; 7-seg display.
                            ; Entering number > binary 1111 terminates
                            ; the program
                            ; ------------------------------------------------

                            ; define all constants used in the program

                  ppi        EQU      0ff40h          ; base address of 8255 PPI
                  switches   EQU      ppi             ; Port A = switches
                  display_led EQU     ppi+1           ; Port B = 7-seg displays
                  control_reg EQU     ppi+3           ; 8255 PPI control register
                  io_defn    EQU      90h             ; I/O mode select byte
                  too_big    EQU      16              ; maximum valid number +1
                  ten        EQU      10              ; first hex number above 9
                  plus_six   EQU      6               ; decimal adjust factor
                  monitor    EQU      0a2h            ; warm restart

8100                        ORG      8100h           ; start of user RAM

                            ; configure 8255 PPI, input Port A, output Port B

8100  90 FF 43  start:     mov      dptr,#control_reg ; set pointer to PPI
8103  74 90                mov      a,#io_defn      ; select operating mode
8105  F0                   movx     @dptr,a         ; and tell PPI

                            ; read binary input switches

8106  90 FF 40  convert:   mov      dptr#switches   ; set pointer to switches
8109  E0                   movx     a,@dptr         ; get input number

                            ; check if binary number greater than 1111, i.e. exit?

810A  B4 10 00             cjne     a,#too_big,c1   ; finished ?
810D  50 0B     c1:        jnc      exit            ; if so then quit

                            ; check if number greater than 9, if so, decimal adjust

810F  B4 0A 00             cjne     a,#ten,c2       ; need correcting?
8112  40 02     c2:        jc       below_ten

                            ; decimal adjust number in range 1010 to 1111
                            ; and display the result

8114  24 06                add      a,#plus_six     ; correction
8116  A3        below_ten: inc      dptr            ; move pointer to leds
8117  F0                   movx     @dptr,a         ; show result

                            ; repeat while input not greater than 1111
                            ; otherwise exit program

8118  80 EC                sjmp     convert         ; iterate while < 16
811A  02 00 A2  exit:      ljmp     monitor         ; return to monitor prog

                            END      start
```

Testing

The design of the program may be checked with appropriate test data, processing by hand as indicated in the structure chart, and recording conditions and actions in tabular form. Complete testing would involve testing with all possible input data values; however, this is time consuming and cannot usually be justified. This program may be tested with just three different values:

1 a value in the range 0000_2 to 1001_2 which requires **no adjustment**,
2 a value in the range 1010_2 to 1111_2 which **does require adjustment**, and
3 a value greater than 1111_2 which **terminates** the program.

Checking with inputs 00000101_2, 00001100_2 and 00010010_2 gives the results shown in Table 5.1 which indicate that the program design is satisfactory.

Table 5.1

Action or condition	State	Input value	Output value	Check OK
1				
3		00000101		
6				
C1	TRUE			
7				
C2	FALSE			
4			00000101	✓
3		00001100		
6				
C1	TRUE			
7				
C2	TRUE			
5				
4			00010010	✓
3		00010010		
6				
C1	FALSE			
2				✓

The program code may be checked in a similar manner by the construction of a **trace table** which shows predictions of the contents of all relevant registers, memory locations and flag conditions. A trace table for the Z80 code is shown in Table 5.2. Similar tables may be constructed for the 8086 and 8031 programs.

Table 5.2

Address bus	Instruction mnemonic		Registers A	Flags C S V Z	PC
			– –	– – – –	1800
1800	LD	A, 90	90	– – – –	1802
1802	OUT	(13),A	90	– – – –	1804
1804	IN	A, (10)	05	– – – –	1806
1806	CP	10	05	1 1 0 0	1808
1808	JP	NC, 1817	05	1 1 0 0	180B
180B	CP	0A	05	1 1 0 0	180D
180D	JP	C, 1812	05	1 1 0 0	1812
1812	OUT	(11),A	05	1 1 0 0	1814
1814	JP	1804	05	1 1 0 0	1804
1804	IN	A, (10)	0C	1 1 0 0	1806
1806	CP	10	0C	1 1 0 0	1808
1808	JP	NC, 1817	0C	1 1 0 0	180B
180B	CP	0A	0C	0 0 0 0	180D
180D	JP	C, 1812	0C	0 0 0 0	1810
1810	ADD	A, 06	12	0 0 0 0	1812
1812	OUT	(11),A	12	0 0 0 0	1814
1814	JP	1804	12	0 0 0 0	1804
1804	IN	A, (10)	18	0 0 0 0	1806
1806	CP	10	18	0 0 0 0	1808
1808	JP	NC, 1817	18	0 0 0 0	180B
1817	RST	30	18	0 0 0 0	180D

After completing a trace table, the relevant machine code may be entered into the target system RAM and executed one step at a time (*single stepped* or *traced*). The state of registers, memory locations and flags may be compared with the predicted results in the trace table. Any discrepancies may then be investigated, and if necessary, adjustments made to the program code.

Test your knowledge 5.1

1   Block X in the JSP chart shown in Figure 5.19 represents:
    A   a sequence
    B   a selection
    C   an iteration
    D   an elementary component

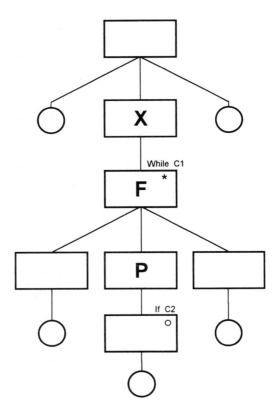

Figure 5.19

2   Block P in the JSP chart shown in Figure 5.19
    represents:
    A   a sequence
    B   a selection
    C   an iteration
    D   an elementary component

3   Block F in the JSP chart shown in Figure 5.19
    represents:
    A   a sequence
    B   a selection
    C   an iteration
    D   an elementary component

4   An 8086 assembly language program contains the
    following line:

```
repeat:    mov    cl,counter  ; reload counter register
```

The underlined part represents:

A   a comment
B   an operand
C   a mnemonic
D   a label

5   A Z80 assembly language program contains the following lines:

```
num1      EQU    00001001b
num2      EQU    30h
total:    ld     a,num1+num2
```

When the program is assembled and executed, the accumulator will be loaded with:

A   30001001b
B   00001032b
C   00111001b
D   00001032h

6   An 8086 assembly language program contains the following lines:

```
1   num1    EQU    135h
2   start:   mov al,num1    ; get  first  number
```

At assembly time an error is reported:

A   in line 1 because 'num1' is not followed by a colon (:)
B   in line 2 because it should be 'mov num1,al'
C   in line 1 because 135 is a decimal number
D   in line 2 because 'num1' is too large for the 'al' register

---

Activity

Refer back to Figure 5.13.
    A program is required to carry out the following operations:

1   read a binary number from the switches,

2   write the number to the 7-segment displays if not greater than 1001_2 (9_{10}),

3 blank the displays by writing FF to the LEDs if a number 1010_2 (10_{10}) to 1111_2 (15_{10}) is entered,

4 terminate the program if a binary input greater than 1111_2 (15_{10}) is entered.

Using the worked example as a guide:

1 Produce preliminary version of program structure diagram.

2 Identify and list conditions and actions.

3 Allocate actions and conditions to program structure diagram.

4 Produce final version of program structure diagram.

5 Write pseudo-code for the program.

6 For selected MPU, write assembly language program and assemble it.

7 Check the program for correct operation by constructing trace tables.

8 Load program code into selected MPU system for final testing.

6 Assembly language programs

Summary

The basic principles of structured programming were dealt with in the previous chapter. This element applies those principles to develop sample software for the Z80, 8086 and 8031 MPUs, and includes programs for performing binary arithmetic, binary to BCD conversion, bit masking, waiting for inputs to change, pulse counting, sequencing, waveform generation, display control and serial I/O. This covers many of the basic operations needed in engineering software, and more complex programs may be formed by combinations of these. JSP structure diagrams and pseudo-code are given for all examples, therefore it should not be difficult to create assembly language programs for MPUs other than those already covered.

Programming examples

Binary addition

Basic arithmetic operations, such as add or subtract, are often needed in control system programs, and these are relatively simple to implement. Such operations may take place with **pure binary** or **BCD numbers**, including or excluding a carry (or a borrow), and with numbers of 8, 16, 32 or more bits (*multiple precision arithmetic*). The following examples demonstrate some of these techniques.

Example 1 – 8-bit binary addition

> A program to perform a binary addition (without carry in) between two 8-bit numbers stored in memory and store the 8-bit sum back into memory (any three appropriate consecutive addresses are used). The sum should always be less than FFH, but if it does exceed this size, then the C-Flag is set, but no other action is expected.

An actions list, structure chart and memory layout for this problem is shown in Figure 6.1.

It can be seen that this is a simple sequence, which may be represented by the following pseudo-code.

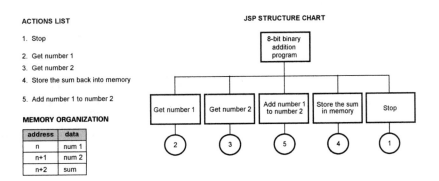

ACTIONS LIST

1. Stop

2. Get number 1

3. Get number 2

4. Store the sum back into memory

5. Add number 1 to number 2

MEMORY ORGANIZATION

address	data
n	num 1
n+1	num 2
n+2	sum

Figure 6.1

8-BIT BINARY ADDITION PROGRAM   *seq*

      DO   2   {get number 1}

      DO   3   {get number 2}

      DO   5   {add number 1 to number 2}

      DO   4   {store the sum back into memory}

      DO   1   {stop}

8-BIT BINARY ADDITION PROGRAM   *end*

The pseudo-code may now be converted into assembly language, and programs for Z80, 8086 and 8031 MPUs are as follows:

Z80

```
             ; ***** Z80 8-bit binary addition **********
             ; PROGRAM 4
             ; performs binary addition using data
             ; stored in memory locations 1900h and
             ; 1901h, and stores the sum in 1902h
             ; ******************************************
             ;
1900 =           num1    EQU     1900h            ; location of number 1
1901 =           num2    EQU     1901h            ; location of number 2
1902 =           sum     EQU     1902h            ; location for sum
                         ;
1800                     ORG     1800h            ; start of user RAM
                         ;
1800 3A 00 19   start:   ld      a,(num1)         ; get number 1
1803 47                  ld      b,a              ; temporary store it
1807 80                  add     a,b              ; add number 1 to it
1808 32 02 19            ld      (sum),a          ; save the sum
                         ;
180B F7                  rst     30h              ; program end
                         ;
                         END     start
```

8086

```
                             ; ***** 8086 8-bit binary addition ******
                             ; PROGRAM 5
                             ; performs a binary addition using data
                             ; stored in locations 0 and 1 of the
                             ; data segment, and stores the sum in
                             ; location 2
                             ; ****************************************
                             ;
                             .MODEL      SMALL
                             .STACK
                             .DATA
                             ;
0000  00        num1         DB          0              ; location of number 1
0001  00        num2         DB          0              ; location of number 2
0002  00        sum          DB          0              ; location for sum
                             ;
                             .CODE
0400                         ORG         0400h          ; start of user RAM
                             ;
0400  B8 ---- R  start:      mov         ax,@DATA       ; set up data segment
0403  8E D8                  mov         ds,ax          ; as that containing num1

0405  A0 0000 R              mov         al,[num1]      ; get number 1
0408  8A 26 0001 R           mov         ah,[num2]      ; get number 2
040C  02 C4                  add         al,ah          ; binary addition
040E  A2 0002 R              mov         [sum],al       ; save the sum
                             ;
0411  CD 20                  int         20h            ; program end
                             ;
                             END         start
```

8031

```
                             ; ***** 8031 8-bit binary addition ******
                             ; PROGRAM 6
                             ; performs a binary addition using data
                             ; stored in internal memory locations
                             ; 40h and 41h, and stores the sum in 42h
                             ; ****************************************
                             ;
                num1         EQU         40h            ; location of number 1
                num2         EQU         41h            ; location of number 2
                sum          EQU         42h            ; location for sum
                monitor      EQU         0a2h           ; warm restart
                             ;
8100                         ORG         8100h          ; start of user RAM
                             ;
8100  E5 40     start:       mov         a,40h          ; get number 1
8102  25 41                  add         a,41h          ; add number 2 to number 1
8104  F5 42                  mov         42h,a          ; save the sum
8106                         ;
8106  02 00 A2               ljmp        monitor        ; program end
                             ;
                             END         start
```

Note: the 8086 program requires **linking** to assign actual addresses to data storage locations (*marked 'R' in the listing*), and these will normally be placed in the next segment following the program code. The 8086 separates **code** and **data** by using **separate segments** which start on any 16-byte boundary throughout the memory map. If data is to be accessed by a program, a data segment pointer (DS) must be set up, and the first two instructions in the program perform this task.

These programs use simple **add** instructions, but some MPUs have only **add with carry** instructions. In such cases a **clear carry** instruction must be used prior to the addition. BCD addition may be achieved by including a **decimal adjust** (DAA) instruction after the add operation, but care must be taken to ensure that numbers 1 and 2 are BCD and do not contain any of the hex characters A to F.

Activity

1   Rewrite Example 1 so that register indirect addressing is used.

2   Devise action lists, structure charts, pseudo-code and program code for:
    (a)   8-bit binary subtraction.
    (b)   8-bit BCD addition.

3   Select suitable test data and construct a trace table for each of these arithmetic programs.

4   Using an appropriate microcomputer system, test each program, check for correct operation and hence verify the trace tables.

Example 2 – 16-bit binary addition

A program to perform a binary addition (without carry in) between two 16-bit numbers stored in memory and store the 16-bit sum back into memory (any six appropriate consecutive addresses may be used). The sum should always be less than FFFFH, but if it does exceed this size, then the C-Flag will be set, but no other action is expected.

An action list, structure chart and memory layout for this problem is shown in Figure 6.2.

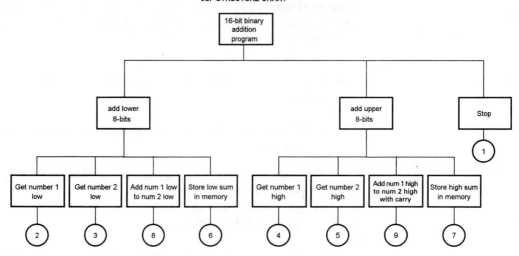

ACTIONS LIST

1. Stop
2. Get number 1 low
3. Get number 2 low
4. Get number 1 high
5. Get number 2 high
6. Store the sum low back into memory
7. Store the sum high back into memory
8. Add number 1 low to number 2 low
9. Add with carry number 1 high to number 2 high

MEMORY ORGANIZATION

address	data
n	num 1 high
n+1	num 1 low
n+2	num 2 high
n+3	num 2 low
n+4	sum high
n+5	sum low

Figure 6.2

This is again a simple sequence with the following pseudo-code:

```
16-BIT BINARY ADDITION PROGRAM    seq
        DO    2        {get number 1 low}
        DO    3        {get number 2 low}
        DO    8        {add number 1 low to number 2 low}
        DO    6        {store the low sum back into memory}
        DO    3        {get number 1 high}
        DO    5        {get number 2 high}
        DO    9        {add number 1 high to number 2 high}
        DO    7        {store the high sum back into memory}
        DO    1        {stop}
16-BIT BINARY ADDITION PROGRAM    end
```

Although it is possible to perform this addition with one operation using 16-bit registers in some cases, separate 8-bit additions have been used to demonstrate the principle of multiple precision arithmetic (arithmetic involving numbers larger than those which can be processed by a single arithmetic instruction). Assembly language programs for the Z80, 8086 and 8031 MPUs are as follows:

Z80

```
                          ; ***** Z80 16-bit binary addition *******
                          ; PROGRAM 7
                          ; performs a binary addition using data
                          ; stored in memory locations 1900h to
                          ; 1903h, and stores the sum in 1904h/1905h
                          ; ***************************************
                          ;
1900 =        num1        EQU        1900h              ; number 1 high byte
1902 =        num2        EQU        1902h              ; number 2 high byte
1904 =        sum         EQU        1904h              ; sum high byte
                          ;
1800                      ORG        1800h              ; start of user RAM
                          ;
                          ; add the low bytes of each number
                          ;
1800 3A 01 19 start:      ld         a,(num1+1)         ; get number 1 low byte
1803 47                   ld         b,a                ; temporary store it
1804 3A 03 19             ld         a,(num2+1)         ; get number 2 low byte
1807 80                   add        a,b                ; add number 1 low byte
1808 32 05 19             ld         (sum+1),a          ; save the sum low byte
                          ;
                          ; add the high bytes of each number together
                          ; with any carry from the previous addition
                          ;
180B 3A 00 19             ld         a,(num1)           ; get number 1 high byte
180E 47                   ld         b,a                ; temporary store it
180F 3A 02 19             ld         a,(num2)           ; get number 2 high byte
1812 88                   adc        a,b                ; add number 1 high byte
1813 32 04 19             ld         (sum),a            ; save the sum high byte
                          ;
1816 F7                   rst        30h                ; program end
                          ;
                          END        start
```

8086

```
                          ; ***** 8086 16-bit binary addition *******
                          ; PROGRAM 8
                          ; performs a binary addition using data
                          ; stored in locations 0 to 3 of the
                          ; data segment, and stores the sum in
                          ; locations 4 and 5
                          ; ***************************************
                          ;
                          .MODEL     SMALL
                          .STACK
                          .DATA
                          ;
0000 0002[    num1        DB         2 dup(0)           ; location of 16-bit number 1
     00
   ]

0002 0002[    num2        DB         2 dup(0)           ; location of 16-bit number 2
     00
   ]
```

```
0004  0002[       sum        DB         2 dup(0)        ; location for 16-bit sum
      00
              ]
                                        ;
                                        .CODE
0400                           ORG        0400h           ; start of user RAM
                                        ;
0400  B8 ---- R   start:      mov        ax,@DATA        ; set up data segment
0403  8E D8                   mov        ds,ax           ; as that containing num1
                                        ;
                                        ; add low bytes of number 1 and number 2
                                        ;
0405  A0 0001 R              mov        al,[num1+1]     ; get number 1 low byte
0408  8A 26 0003 R           mov        ah,[num2+1]     ; get number 2 low byte
040C  02 C4                  add        al,ah           ; binary addition
040E  A2 0005 R              mov        [sum+1],al      ; save low byte of sum
                                        ;
                                        ; add high bytes of number 1 and number 2 plus
                                        ; any carry from the previous addition
                                        ;
0411  A0 0000 R              mov        al,[num1]       ; get number 1 high byte
0414  8A 26 0002 R           mov        ah,[num2]       ; get number 2 high byte
0418  12 C4                  adc        al,ah           ; binary addition
041A  A2 0004 R              mov        [sum],al        ; save high byte of sum
                                        ;
041D  CD 20                  int        20h             ; program
                                        ;
                              END        start
```

8031

```
                                        ; ***** 8031 16-bit binary addition ******
                                        ; PROGRAM 9
                                        ; performs a binary addition using data
                                        ; stored in internal memory locations
                                        ; 40h to 43h, and store the sum in 44h/45h
                                        ; ****************************************
                                        ;
                    num1       EQU        40h             ; number 1 high byte
                    num2       EQU        42h             ; number 2 high byte
                    sum        EQU        44h             ; sum high byte
                    monitor    EQU        0a2h            ; warm restart

                                        ;
8100                           ORG        8100h           ; start of user RAM
                                        ;
                                        ; add number 1 and number 2 low bytes
                                        ;
8100  E5 41      start:      mov        a,num1+1        ; get number 1 low byte
8102  25 43                  add        a,num2+1        ; add number 2 low byte
8104  F5 45                  mov        sum+1,a         ; save low byte of sum
                                        ;
                                        ; add number 1 and number 2 high bytes plus
                                        ; any carry from the previous addition
                                        ;
8106  E5 40                  mov        a,num1          ; get number 1 low byte
8108  35 42                  addc       a,num2          ; add number 2 to number 1
810A  F5 44                  mov        sum,a           ; save the sum
                                        ;
810C  02 00 A2               ljmp       monitor         ; program
                                        ;
```

Note that the first addition is performed using an ordinary **add** instruction, but the second addition uses an **add with carry** instruction. Further use of the 'add with carry' instruction may be applied to subsequent pairs of numbers so that there is virtually no limit to the sizes of numbers that may be added.

Activity

1   Rewrite Example 2 so that a single 16-bit add operation is used (Z80 and 8086 MPUs only).

2   Devise action lists, structure charts, pseudo-code and program code for:
(a)   16-bit BCD addition,
(b)   16-bit subtraction.

3   Select suitable test data and construct a trace table for each of these arithmetic programs.

4   Using an appropriate microcomputer system, test each program, check for correct operation and hence verify the trace tables.

Multiplication

Many microprocessors now include a **multiply** instruction in their instruction set, but if this is not available it is possible to perform multiplication by a number of different algorithms. One simple algorithm uses repeated addition, and programs for this method are shown in Examples 3 and 4.

Example 3 – Multiplication by 5

An interface circuit of the type shown in Figure 5.13 is connected to a microcomputer. A program is required to read the switches connected to Port A, multiply the reading by five and display the BCD result on the 7-segment displays connected to Port B. If a number greater than 9 (00001001_2) is entered then the program is terminated.

Multiplication by a constant such as five may be achieved by a few simple additions, i.e. add the input number to itself to multiply by two, add the result of this to itself to multiply by four, then add in the original number. The structure chart for this process is shown in Figure 6.3.

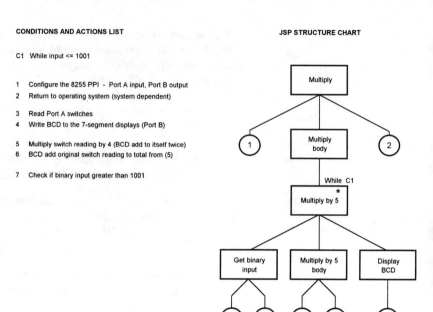

CONDITIONS AND ACTIONS LIST

C1 While input <= 1001

1 Configure the 8255 PPI - Port A input, Port B output
2 Return to operating system (system dependent)

3 Read Port A switches
4 Write BCD to the 7-segment displays (Port B)

5 Multiply switch reading by 4 (BCD add to itself twice)
6 BCD add original switch reading to total from (5)

7 Check if binary input greater than 1001

Figure 6.3

Pseudo-code for this program is as follows:

```
MULTIPLY  seq
    DO    1                  {configure 8255 PPI}
    MULTIPLY BODY iter WHILE C1   {input number <= 9}
        MULTIPLY BY 5 seq
            DO    3          {read switches}
            DO    7          {check if input > 9}
            DO    5          {multiply input by 4}
            DO    6          {add original switch input}
            DO    4          {write BCD to 7-seg displays}
        MULTIPLY BY 5 end
    MULTIPLY BODY end
    DO    2                  {return to operating system}
MULTIPLY  end
```

The following listings show typical assembly language programs for the Z80, 8086 and 8031 MPUs.

Z80

```
                        ; *********** Z80 multiply by 5 ************
                        ; PROGRAM 10
                        ; multiplies a BCD inpute from port switches
                        ; by 5 and displays the BCD result on
                        ; 7-segment LEDs.
                        ; Terminates if non-BCD input occurs.
                        ; Uses circuit shown in Figure 5.13
                        ; *******************************************
                        ;
0010 =      ppi         EQU         10h             ; 8255 PPI base address
0010 =      switch      EQU         ppi             ; switches (Port A)
0011 =      leds        EQU         ppi+1           ; 7-seg displays (Port B)
0013 =      control     EQU         ppi+3           ; PPI control register
0090 =      iobyte      EQU         90h             ; I/O definition byte
000A =      too_big     EQU         10              ; BCD is 0 to 9 only
                        ;
1800                    ORG         1800h           ; start of user RAM
                        ;
                        ; configure the 8255 PPI ports
                        ;
1800 3E 90  start:      ld          a, iobyte       ; set up Port A as inputs
1802 D3 13              out         (control),a     ; and Port B as outputs
                        ;
                        ; check for non-BCD input
                        ;
1804 DB 10  conv:       in          a,(switch)      ; read switches
1806 FE 0A              cp          too_big         ; check if > 9
1808 D2 17 18           jp          nc,exit         ; if so then quit
                        ;
                        ; multiply by 5 using double and add method
                        ;
180B 47     mult:       ld          b,a             ; save input (= acc x 1)
180C 87                 add         a,a             ; double acc (= acc x 2)
180D 27                 daa                         ; keep result as BCD
180E 87                 add         a,a             ; double acc (= acc x 4)
180F 27                 daa
1810 80                 add         a,b             ; add input (= acc x 5)
1811 27                 daa
                        ;
                        ; display result on the 7-segment displays
                        ;
1812 D3 11              out         (leds),a        ; show result
1814 C3 04 18           jp          conv            ; and repeat
                        ;
                        ; exit if non-BCD input applied
                        ;
1817 F7     exit:       rst         30h             ; return to monitor program
                        ;
                        END         start
```

8086

```
                    ; *********** 8086 multiply by 5 ************
                    ; PROGRAM 11
                    ; multiplies a BCD inpute from port switches
                    ; by 5 and displays the BCD result on
                    ; 7-segment LEDs.
                    ; Terminates if non-BCD input occurs.
                    ; Uses circuit shown in Figure 5.13
                    ; ******************************************
                    ;
                    .MODEL      SMALL
                    .STACK
                    ;
=      0010    ppi      EQU      10h          ; 8255 PPI base address
=              switch   EQU      ppi          ; switches (Port A)
=      0011    leds     EQU      ppi+1        ; 7-seg displays (Port B)
=      0013    control  EQU      ppi+3        ; PPI control register
=      0090    iobyte   EQU      90h          ; I/O definition byte
=      000A    too_big  EQU      10           ; BCD is 0 to 9 only
                    ;
                    .CODE
0400                ORG      0400h            ; start of user RAM
                    ;
                    ; configure the 8255 PPI ports
                    ;
0400 B0 90  start:  mov      al, iobyte       ; set up Port A as inputs
0402 E6 13          out      control,al       ; and Port B as outputs
                    ;
                    ; check for non-BCD input
                    ;
0404 E4 10  conv:   in       al,(switch)      ; read switches
0406 3C 0A          cmp      al,too_big       ; check if > 9
0408 73 0F          jnb      exit             ; if so then quit
                    ;
                    ; multiply by 5 using double and add method
                    ;
040A 8A E0  mult:   mov      ah,al            ; save input (= acc x 1)
040C 02 C0          add      al,al            ; double acc (= acc x 2)
040E 27             daa                       ; keep result as BCD
040F 02 C0          add      al,al            ; double acc (= acc x 4)
0411 27             daa
0412 02 C4          add      al,ah            ; add input (= acc x 5)
1811 27             daa
                    ;
                    ; display result on the 7-segment displays
                    ;
0415 E6 11          out      leds,al          ; show result
0417 EB EB          jmp      conv             ; and repeat
                    ;
                    ; exit if non-BCD input applied
                    ;
0419 CD 20  exit:   int      20h              ; return to monitor program
                    ;
                    END      start
```

8031

```
                    ; *********** 8031 multiply by 5 ***********
                    ; PROGRAM 12
                    ; multiplies a BCD input from port switches
                    ; by 5 and displays the BCD result on
                    ; 7-segment LEDs.
                    ; Terminates if non-BCD input occurs.
                    ; Uses circuit shown in Figure 5.13
                    ; *******************************************
                    ;
                    ;
          ppi       EQU     0ff40h          ; 8255 PPI base address
          switch    EQU     ppi             ; switches (Port A)
          leds      EQU     ppi+1           ; 7-seg displays (Port B)
          control   EQU     ppi+3           ; PPI control register
          iobyte    EQU     90h             ; I/O definition byte
          too_big   EQU     10              ; BCD is 0 to 9 only
          monitor   EQU     0a2h            ; warm restart
                    ;
8100                ORG     8100h           ; start of user RAM
                    ;
                    ; configure the 8255 PPI ports
                    ;
8100 90 FF 43 start: mov    dptr,#control   ; point to PPI control reg
8103 74 90          mov     a,#iobyte       ; set up Port A as inputs
8105 F0             movx    @dptr,a         ; and Port B as outputs
8106                ;
8106                ; check for non-BCD input
8106                ;
8106 90 FF 40 conv: mov     dptr,#switch    ; point to switches
8109 E0             movx    a,@dptr         ; read switches
810A B4 0A 00       cjne    a,#too_big,check ; check if > 9
810D 50 0C  check:  jnc     exit            ; if so then quit
                    ;
                    ; multiply by 5 using double and add method
                    ;
810F FA     mult:   mov     r2,a            ; save input (= acc x 1)
8110 2A             add     a,r2            ; double acc (= acc x 2)
8111 D4             da      a               ; keep result as BCD
8112 FB             mov     r3,a
8113 2B             add     a,r3            ; double acc (= acc x 4)
8114 D4             da      a
8115 2A             add     a,r2            ; add input (= acc x 5)
8116 D4             da      a
                    ;
                    ; display result on the 7-segment displays
                    ;
                    inc     dptr            ; point to leds,
8118 F0             movx    @dptr,a         ; show result
8119 80 EB          sjmp    conv            ; and repeat
                    ;
                    ; exit if non-BCD input applied
                    ;
811B 02 00 A2 exit: ljmp    monitor         ; return to monitor program
                    ;
                    END     start
```

Activity

1   Devise an action list, structure chart, pseudo-code and
program code for a program similar to Example 3 but
for multiplication by seven.

2   Construct a trace table for your program, selecting
suitable input test data.

3   Using an appropriate microcomputer system, test your
program, check for correct operation and hence verify
the trace table.

Multiplication by repeated addition

The previous example illustrated how multiplication by a constant may be
achieved. When the multiplier is a variable, repeated addition may be used
and this requires the multiplier to be used as a **loop counter** which
determines how many times the multiplicand is added to a running total.
When any two 8-bit numbers are multiplied, the product may be
considerably larger than eight bits, therefore 16 bits are needed as storage for
the product.

Example 4 – Multiplication of two numbers by repeated addition

A program is required to perform multiplication between two
8-bit binary numbers using a repeated addition technique. The
two numbers are stored in any suitable memory locations and
the 16-bit product should be stored back into memory.

A structure chart for multiplication by repeated addition is shown in
Figure 6.4.

The pseudo-code for this example is as follows:

```
8 x 8 BIT MULTIPLY PROGRAM    seq
        DO  1                   {clear product register}
        DO  2                   {get multiplicand from memory}
        DO  3                   {get multiplier from memory}
        MULTIPLY iter  WHILE C1  {multiplier not zero}
            MULTIPLY BODY seq
                    DO  8       {check multiplier for zero}
                    DO  6       {add multiplicand to product
                                sub-total}
```

$$
\begin{array}{lll}
\text{DO} & 7 & \{\text{decrement multiplier by 1}\}\\
\end{array}
$$

MULTIPLY BODY *end*

MULTIPLY *end*

$$
\begin{array}{lll}
\text{DO} & 5 & \{\text{store product in memory}\}\\
\text{DO} & 4 & \{\text{return to operating system}\}\\
\end{array}
$$

8 x 8 BIT MULTIPLY PROGRAM *end*

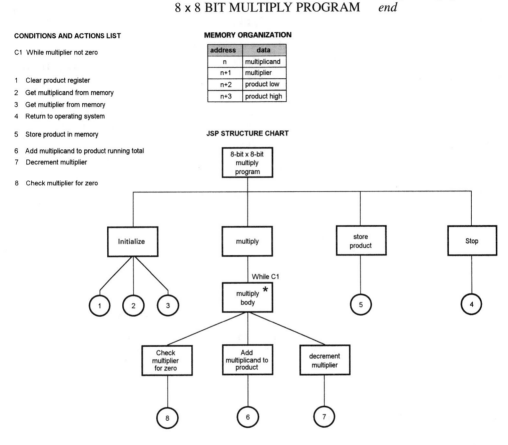

CONDITIONS AND ACTIONS LIST

C1 While multiplier not zero

1 Clear product register
2 Get multiplicand from memory
3 Get multiplier from memory
4 Return to operating system

5 Store product in memory

6 Add multiplicand to product running total
7 Decrement multiplier

8 Check multiplier for zero

MEMORY ORGANIZATION

address	data
n	multiplicand
n+1	multiplier
n+2	product low
n+3	product high

JSP STRUCTURE CHART

Figure 6.4

The following listings show typical assembly language programs for the Z80, 8086 and 8031 MPUs.

Z80

```
; ********** Z80 multiply by repeated addition **
; PROGRAM 13
; performs binary multiplication between
; two numbers stored in memory locations
; 1900h and 1901h using a repeated addition
; method and stores the product in 1902h/1903h
; ***********************************************
;
1900 =          mcand    EQU    1900h        ; number 1 (multiplicand)
1901 =          mplier   EQU    1901h        ; number 2 (multiplier)
1902 =          product  EQU    1902h        ; product low byte
;
```

```
1800                            ORG      1800h              ; start of user RAM
                                ;
                                ; initialize registers used for multiplication
                                ;
1800  21 00 00    start:        ld       hl,0               ; clear product register
1803  3A 00 19                  ld       a,(mcand)          ; get multiplicand
1806  5F                        ld       e,a                ; and transfer to adding
1807  55                        ld       d,l                ; register (clear high byte)
1808  3A 01 19                  ld       a,(mplier)         ; get multiplier
                                ;
                                ; check for multiply by zero
                                ;
180B  B7                        or       a                  ; check for multiplier = 0
180C  CA 14 18    check:        jp       z,store            ; store the product
                                ;
                                ; perform repeated addition
                                ;
180F  19                        add      hl,de              ; add multiplicand to total
1810  3D                        dec      a                  ; reduce multiplier by 1
1811  C3 0C 18                  jp       check              ; and repeat
                                ;
                                ; store the 16-bit product
                                ;
1814  22 02 19    store:        ld       (product),hl       ; low byte first
                                ;
1817  F7                        rst      30h                ; program end
                                ;
                                END      start
```

8086

```
                   ; ****** 8086 multiply by repeated addition ******
                   ; PROGRAM 14
                   ; performs binary multiplication between
                   ; two numbers stored in data segment locations
                   ; 0 and 1 using a repeated addition method and
                   ; stores the product in locations 2 and 3
                   ; ************************************************
                   ;
                   ;
                   .MODEL   SMALL
                   .STACK
                   .DATA
                   ;
0000  00           mcand     DB      0              ; number 1 (multiplicand)
0001  00           mplier    DB      0              ; number 2 (multiplier)
0002  0000         product   DW      0              ; product low byte
                             ;
                             .CODE
0400                         ORG     0400h          ; start of user RAM
                             ;
                             ; initialize registers used for multiplication
                             ;
0400  B8 ---- R    start:    mov     ax,@DATA       ; set up data segment
0403  8E D8                  mov     ds,ax          ; as that containing num1
0405  B8 0000                mov     ax,0           ; clear product register
0408  8A 1E 0000 R           mov     bl,[mcand]     ; get multiplicand
040C  8A FC                  mov     bh,ah          ; get multiplicand high
040E  8A 0E 0001 R           mov     cl,[mplier]    ; get multiplier
                             ;
                             ; check for multiply by zero
                             ;
```

```
0412  0A C9                     or      cl,cl              ; check for multiplier = 0
0414  74 06        check:       jz      store              ; store the product
                                ;
                                ; perform repeated addition
                                ;
0416  03 C3                     add     ax,bx              ; add multiplicand to total
0418  FE C9                     dec     cl                 ; reduce multiplier by 1
041A  75 F8                     jmp     check              ; and repeat
                                ;
                                ; store the 16-bit product
                                ;
041C  A3 0002 R    store:       mov     [product],ax       ; store low byte first
                                ;
041F  CD 20                     int     20h                ; program end
                                ;
                                END     start
```

8031

```
                                ; ****** 8031 multiply by repeated addition ******
                                ; PROGRAM 15
                                ; performs a binary multiplication between two
                                ; numbers stored in internal memory locations
                                ; 40h and 41h using a repeated addition
                                ; *************************************************
                                ;
                 mcand    EQU    40h                ; number 1 (multiplicand)
                 mplier   EQU    41h                ; number 2 (multiplier)
                 product  EQU    42h                ; product low byte
                 monitor  EQU    0a2h               ; warm restart
                                ;
8100                            ORG    8100h              ; start of user RAM
                                ;
                                ; initialize registers used for multiplication
                                ;
0019  8100 75 42 00  start:     mov    product,#0         ; clear product low
0020  8103 75 43 00             mov    product+1,#0       ; clear product high
0021  8106 78 42                mov    r0,#product        ; pointer product low
0022  8108 79 40                mov    r1,#mcand          ; pointer multiplicand
                                ;
                                ; check for multiply by zero
                                ;
0026  810A E4                   clr    a                  ; clear accumulator
0027  810B 45 41                orl    a,mplier           ; check for multiplier = 0
0028  810D 60 0A    check:      jz     exit               ; exit if zero
                                ;
                                ; perform repeated addition
                                ;
0032  810F E7                   mov    a,@r1              ; add multiplicand to total
0033  8110 26                   add    a,@r0              ; add product sub total
0034  8111 F6                   mov    @r0,a              ; update subtotal
0035  8112 50 02                jnc    loop               ; no increase in product high
0036  8114 05 43                inc    product+1          ; add 1 to product high
0037  8116 D5 41 F6  loop:      djnz   mplier,check       ; repeat until multiplier = 0
                                ;
0039  8119 02 00 A2  exit:      ljmp   monitor            ; program
                                ;
                                END    start
```

Binary to BCD

Performing arithmetic with BCD numbers, particularly multiplication or division, can be rather difficult for a microprocessor, therefore it is more usual to carry out all calculations in **pure binary**. When displaying results, however, a human operator expects results to be displayed in a readily recognizable form, i.e. **denary**, therefore some form of binary to BCD conversion is required.

The structured programming example shown earlier in this chapter dealt with a simple binary to BCD conversion process (adding 6 if the number entered was greater than 9). This simple method is suitable for single digit input, but becomes more complex if larger numbers are to be converted and an alternative algorithm may be required. For example, an 8-bit binary number ($0–255_{10}$) may be converted to BCD by first dividing by 100_{10} to evaluate the hundreds, dividing the remainder by 10_{10} to evaluate the tens, leaving the remainder as the units. An alternative algorithm which is relatively simple to implement in a microcomputer program is known as the **double and add** method, shown in Figure 6.5.

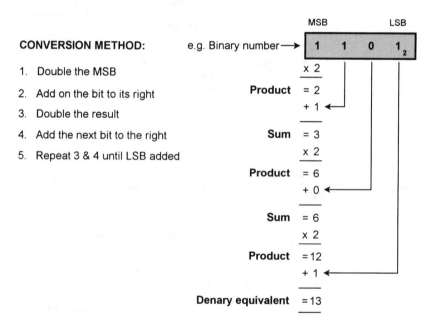

CONVERSION METHOD: e.g. Binary number

1. Double the MSB
2. Add on the bit to its right
3. Double the result
4. Add the next bit to the right
5. Repeat 3 & 4 until LSB added

Figure 6.5

Example 5 – Binary to BCD (double and add method)

An interface circuit of the type shown in Figure 5.13 is connected to a microcomputer. A program is required to read the switches connected to Port A, convert the binary input into its BCD equivalent using a 'double and add' method, and display the result on the 7-segment displays. If a number greater than 01100011_2 (63_{16}) is entered, the program is terminated.

A structure chart for this method is shown in Figure 6.6.

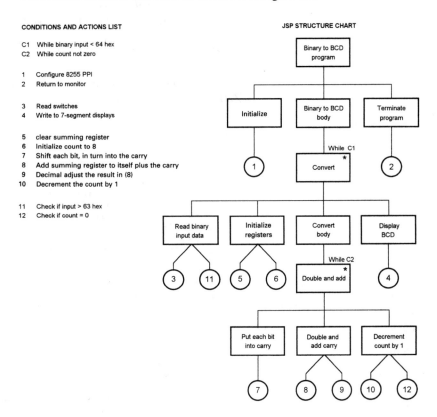

CONDITIONS AND ACTIONS LIST

C1   While binary input < 64 hex
C2   While count not zero

1   Configure 8255 PPI
2   Return to monitor

3   Read switches
4   Write to 7-segment displays

5   clear summing register
6   Initialize count to 8
7   Shift each bit, in turn into the carry
8   Add summing register to itself plus the carry
9   Decimal adjust the result in (8)
10   Decrement the count by 1

11   Check if input > 63 hex
12   Check if count = 0

Figure 6.6

The pseudo-code for this example is as follows:

```
BINARY TO BCD  PROGRAM    seq
    DO  1                          {configure 8255 PPI}
        BINARY TO BCD BODY iter  WHILE C1 {input < 64 hex}
        CONVERT seq
            DO   3                 {read switches}
            DO   11                {check if input > 63 hex}
            DO   5                 {clear summing register}
            DO   6                 {initialize count to 8}
            CONVERT BODY iter   WHILE C2  {count not zero}
                DOUBLE AND ADD  seq
                    DO  7     {shift each bit into carry}
```

DO 8	{add sum reg to itself + carry}
DO 9	{decimal adjust result}
DO 10	{decrement count by 1}
DO 12	{check if count = 0}

DOUBLE AND ADD   *end*

CONVERT BODY *end*

DO 4   {write to 7-seg displays}

CONVERT *end*

BINARY TO BCD BODY   *end*

DO 2   {return to monitor}

BINARY TO BCD PROGRAM   *end*

The following listings show typical assembly language programs for the Z80, 8086 and 8031 MPUs.

Z80

```
        ; *********** Z80 binary to BCD ***********
        ; PROGRAM 16
        ; converts binary input from port switches
        ; into BCD and displays the result on
        ; two-digit 7-segment LEDs.
        ; Terminates if number > 99 is entered.
        ; Uses circuit shown in Figure 5.13
        ; ********************************************
        ;
0010 =          ppi         EQU     10h             ; 8255 PPI base address
0010 =          switch      EQU     ppi             ; switches (Port A)
0011 =          leds        EQU     ppi+1           ; 7-seg displays (Port B)
0013 =          control     EQU     ppi+3           ; PPI control register
0090 =          iobyte      EQU     90h             ; I/O definition byte
0064 =          too_big     EQU     100             ; 0 to 99 only
                            ;
1800                        ORG     1800h           ; start of user RAM
                            ;
                            ; configure the 8255 PPI ports
                            ;
1800 3E 90      start:      ld      a, iobyte       ; set up Port A as inputs
1802 D3 13                  out     (control),a     ; and Port B as outputs
                            ;
                            ; check for input number greater than 99
                            ;
1804 D8 10      conv:       in      a,(switch)      ; read switches
1806 FE 64                  cp      too_big         ; check if > 99
1808 D2 1A 18               jp      nc,exit         ; if so then quit
                            ;
                            ; convert binary input to BCD
                            ;
180B 06 08                  ld      b,8             ; count 8 double and adds
180D 4F                     ld      c,a             ; copy binary input to C reg
180E AF                     xor     a               ; clear accumulator
180F CB 11     shift:       rl      c               ; shift bits into C flag
1811 8F                     adc     a,a             ; double and add
1812 27                     daa                     ; keep as BCD
1813 10 FA                  djnz    shift           ; repeat for all 8 bits
                            ;
                            ; display result on the 7-segment displays
                            ;
1815 D3 11                  out     (leds),a        ; show result
```

```
1817  C3 04 18                 jp        conv          ; and repeat
                               ;
                               ; exit if non-BCD input applied
                               ;
181A  F7         exit:         rst       30h           ; return to monitor program
                               ;
                               END       start
```

8086

```
                     ; ********** 8086 binary to BCD **********
                     ; PROGRAM 17
                     ; converts binary input from port switches
                     ; into BCD and displays the result on
                     ; two-digit 7-segment LEDs.
                     ; Terminates if number > 99 is entered.
                     ; Uses circuit shown in Figure 5.13
                     ; ****************************************
                     ;
                     .MODEL    SMALL
                     .STACK
                     ;
= 0010     ppi       EQU       10h           ; 8255 PPI base address
=          switch    EQU       ppi           ; switches (Port A)
= 0011     leds      EQU       ppi+1         ; 7-seg displays (Port B)
= 0013     control   EQU       ppi+3         ; PPI control register
= 0090     iobyte    EQU       90h           ; I/O definition byte
= 0064     too_big   EQU       100           ; 0 to 99 only
                     ;
                     .CODE
0400                 ORG       0400h         ; start of user RAM
                     ;
                     ; configure the 8255 PPI ports
                     ;
0400  B0 90   start: mov       al,iobyte     ; set up Port A as inputs
0402  E6 13          out       (control),al  ; and Port B as outputs
                     ;
                     ; check for input number greater than 99
                     ;
0404  E4 10   conv:  in        al,(switch)   ; read switches
0406  3C 64          cmp       al,too_big    ; check if > 99
0408  73 12          jnp       exit          ; if so then quit
                     ;
                     ; convert binary input to BCD
                     ;
040A  B9 0008        mov       cx,8          ; count 8 double and adds
040D  8A D8          mov       bl,al         ; copy binary number to BL
040F  32 C0          xor       al,al         ; clear AL register
0411  D0 E3   shift: shl       bl,1          : shift bits into C flag
0413  12 C0          adc       al,al         ; double and add
0415  27             daa                     ; keep as BCD
0416  E2 F9          loop      shift         ; repeat for all 8 bits
                     ;
                     ; display result on the 7-segment displays
                     ;
0418  E6 11          out       leds,al       ; show result
041A  7A E8          jp        conv          ; and repeat
                     ;
                     ; exit if input > 99 entered
                     ;
041C  CD 20   exit:  int       20h           ; return to monitor program
                     ;
                     END       start
```

8031

```
                          ; ********** 8031 binary to BCD **********
                          ; PROGRAM 18
                          ; converts binary input from port switches
                          ; into BCD and displays the result on
                          ; two-digit 7-segment LEDs.
                          ; Terminates if number > 99 is entered.
                          ; Uses circuit shown in Figure 5.13
                          ; ****************************************
                          ;
              ppi     EQU       0ff40h          ; 8255 PPi base address
              switch  EQU       ppi             ; switches (Port A)
              control EQU       ppi+3           ; PPI control register
              iobyte  EQU       90h             ; I/O definition byte
              too_big EQU       100             ; 0 to 99 only
              monitor EQU       0a2h            ; warm restart
                      ;
8100                  ORG       8100h           ; start of user RAM
                      ;
                      ; configure the 8255 PPI ports
                      ;
8100 90 FF 43 start:  mov       dptr,#control   ; point to PPI control reg
8103 74 90            mov       a,#iobyte       ; set up Port A as inputs
8105 F0               movx      @dptr,a         ; and Port B as outputs
                      ;
                      ; check for input number greater than 99
                      ;
8106 90 FF 40 conv:   mov       dptr,#switch    ; point to switches
8109 E0               movx      a,@dptr         ; read switches
810A B4 64 00         cjne      a,#too_big,check ; check if > 99
810D 50 10    check:  jnc       exit            ; if so then quit
                      ;
                      ; convert binary input to BCD
                      ;
810F 7A 08            mov       r2,#8           ; count 8 double and adds
8111 FB               mov       r3,a            ; copy binary input to R3 reg
8112 E4               clr       a               ; clear accumulator
8113 FC       shift:  mov       r4,a            ; doubling register
8114 CB               xch       a,r3            ; borrow acc for the shift
8115 33               rlc       a               ; shift bits into C flag
8116 CB               xch       a,r3            ; swap data back
8117 3C               addc      a,r4            ; double and add
8118 D4               da        a               ; keep as BCD
8119 DA F8            djnz      r2,shift        ; repeat for all 8 bits
                      ;
                      ; display result on the 7-segment displays
                      ;
0048 811B A3          inc       dptr            ; point to leds,
0049 811C F0          movx      @dptr,a         ; show result
0050 811D 80 E7       sjmp      conv            ; and repeat
                      ;
                      ; exit if non-BCD input applied
                      ;
811F 02 00 A2 exit:   ljmp      monitor         ; return to monitor program
                      ;
              END     start
```

Activity

1   Devise an **action list**, **structure chart**, **pseudo-code** and **program code** for a program similar to that in Example 5 but which does its conversion by repeated subtraction of 10.

2   Construct a **trace table** for your program, selecting suitable input test data.

3   Using an appropriate microcomputer system, **test your program**, check for correct operation and hence verify the trace table.

Time delays

Microcomputers operate at very high speeds, performing millions of instructions every second. While this is advantageous for many applications, it is a distinct disadvantage when controlling equipment that responds at much slower rates, for example mechanical equipment or human operators. Such applications require **delay loops** to be built into programs which have the effect of making a computer wait between operations.

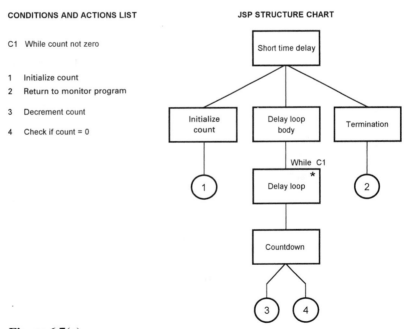

CONDITIONS AND ACTIONS LIST

C1   While count not zero

1   Initialize count
2   Return to monitor program

3   Decrement count

4   Check if count = 0

JSP STRUCTURE CHART

Figure 6.7(a)

The usual way of implementing delay loops is to make a microcomputer count down to zero from a very high number. Each MPU instruction requires a certain number of clock cycles for its execution, and since each clock cycle occupies a fixed time period, its execution time may be determined. Although this time is very short, if the same instructions are repeated millions of times an appreciable time delay results.

The operation of a time delay loop may be studied by referring to the structure charts shown in Figures 6.7(a) and (b).

The maximum number that can be used for a count down depends upon the size of register available, i.e. 255 for an 8-bit register and 65 535 for a 16-bit register. This may be insufficient for some applications, and in such cases it may be necessary to put one delay loop inside of another delay loop (nested loops) as shown in Figure 6.7(b).

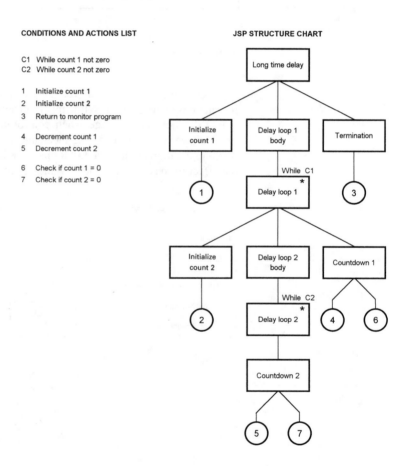

Figure 6.7(b)

Pseudo-code for the time delays shown in Figures 6.7(a) and (b) are as follows:

```
SHORT TIME DELAY  seq
    DO    1                                {initialize count}
    DELAY LOOP BODY  iter       WHILE C1    {count not zero}
        DELAY LOOP  seq
            DO    3                         {decrement count by 1}
            DO    4                         {check for count = 0}
        DELAY LOOP  end
    DELAY LOOP BODY  end
    DO    2                                 {return to monitor program}
SHORT TIME DELAY  end


LONG TIME DELAY  seq
    DO    1                                 {initialize count1}
    DELAY LOOP 1 BODY  iter   WHILE C1 {count1 not zero}
        DELAY LOOP 1  seq
            DO    2             {initialize count 2}
            DELAY LOOP 2 BODY  iter WHILE C2 {count2 <>0}
                DELAY LOOP 2  seq
                    DO    5   {decrement count 2 by 1}
                    DO    7   {check for count 2 = 0}
                DELAY LOOP 2  end
            DELAY LOOP 2 BODY  end
            DO    4             {decrement count 1 by 1}
            DO    6             {check for count 1 = 0}
        DELAY LOOP 1  end
    DELAY LOOP 1 BODY  end
    DO    3                                 {return to monitor program}
LONG TIME DELAY  end
```

The time delay may be calculated by multiplying the total number of cycles in the delay loop by the clock period. The total number of cycles is obtained by adding together the cycles for each instruction, each multiplied by the number of times it is executed in the loop. As an example of this, consider the following 8086 MPU delay loop:

```
0400        mov         cx,0e5c8h
0403        loop        0403h
```

If this loop is executed in a system with a 2 MHz clock, a 500 ms or 0.5 s time delay results, as shown in Table 6.1.

Table 6.1

Instruction	Number of 0.5 µs cycles	Number of times executed	Total cycles
mov cx,0e5c8h	4	1	4
loop 0403h (jump)	17	58 823	999 991
loop 0403h (no jump)	5	1	5
Total clock cycles for program			1 000 000
Total execution (1 000 000 x 0.5 µs)			0.5 s

Example 6 – Flashing LED

A microcomputer has a single LED connected to Port B of an
8255 PPI as shown in Figure 6.8. A program is required to cause
the LED to be ON for 0.5 s and OFF for 0.5 s repetitively, i.e.
flash at a 1 Hz rate, using a timing loop of the type already
described.

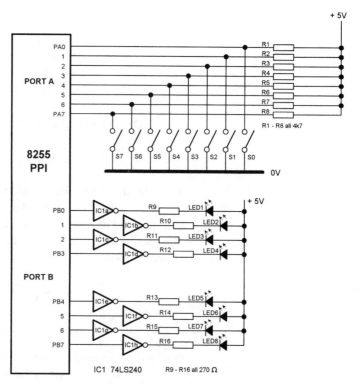

Figure 6.8

A single loop using a 16-bit counter may be possible with some systems, but nested loops may be necessary to obtain a long enough time delay. A structure diagram for this system is shown in Figure 6.9.

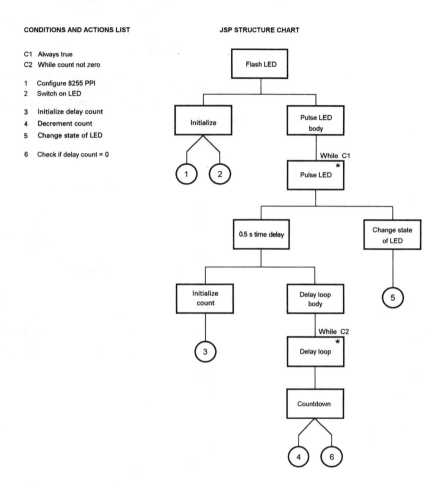

CONDITIONS AND ACTIONS LIST

C1 Always true
C2 While count not zero

1   Configure 8255 PPI
2   Switch on LED

3   Initialize delay count
4   Decrement count
5   Change state of LED

6   Check if delay count = 0

JSP STRUCTURE CHART

Figure 6.9

Pseudo-code for the flashing LED program is as follows:

```
FLASH LED  seq
    DO  1                   {configure 8255 PPI – Port B outputs}
    DO  2                   {switch on LED}
    PULSE LED BODY iter  WHILE C1      {indefinitely}
        0.5 s TIME DELAY  seq
            DO  3          {initialize delay counter}
            DELAY LOOP BODY iter  WHILE C2 {count <> zero}
                DELAY LOOP  seq
                    DO  4     {decrement count by 1}
```

```
                              DO   6    {check for count = 0}
                       DELAY LOOP   end
                  DELAY LOOP BODY   end
              0.5 s TIME DELAY   end
                    DO 5                  {change state of LED}
                PULSE LED BODY   end
            FLASH LED   end
```

The following listings show typical assembly language programs for the Z80, 8086 and 8031 MPUs, and timing calculations are shown in Tables 6.2 (a), (b) and (c) . Note that the 8031 MPU requires nested loops to obtain the correct time delay.

Z80

```
; ********** Z80 Flash LED program **********
; PROGRAM 19
; flashes LED connected to bit 0 of Port B
; ON for 0.5 s and OFF for 0.5 s continuously.
; Uses circuit shown in Figure 6.8.
; Program assumes clock frequency of 2 MHz.
; Time for a 'T' state = 0.5 microseconds
; ***********************************************
;
0010 =          ppi       EQU       10h         ; 8255 PPi base address
0011 =          leds      EQU       ppi+1       ; display (Port B)
0013 =          control   EQU       ppi+3       ; PPI control register
0080 =          iobyte    EQU       80h         ; I/O definition byte
0001 =          on        EQU       1           ; LED ON byte
A2C1 =          h_sec     EQU       41665       ; 1/2 sec loop count
                          ;
1800                      ORG       1800h       ; start of user RAM
                          ;
                          ; configure 8255 PPI
                          ;
1800 3E 80      start:    ld        a,iobyte    ; configure 8255 PPI
1802 D3 13                out       (control),a ; with Port B as outputs
                          ;
                          ; switch on LED
                          ;
1804 3E 01                ld        a,on        ; set bit 0 of Port B
1806 D3 11      flash:    out       (leds),a    ; to switch on LED      (11T)
1808 47                   ld        b,a         ; save LED status        (4T)
                          ;
                          ; 0.5 second time delay (10T = {6T+4T+4T+10T} * 41665)
                          ;
1809 21 C1 A2             ld        hl,h_sec    ; initialize delay count (10T)
180C 2B         delay:    dec       hl          ; count down to zero      (6T)
180D 7C                   ld        a,h         ; but 'dec hl' does not   (4T)
180E B5                   or        l           ; set any flags - use acc (4T)
180F C2 0C 18             jp        nz,delay    ; to test count in hl    (10T)
                          ;
                          ; change state of LED
                          ;
1812 78                   ld        a,b         ; recover LED status      (4T)
1813 EE 01                xor       l           ; invert bit 0            (7T)
1815 C3 06 18             jp        flash       ; and repeat             (10T)
                          ;
                          END       start
```

8086

```
                        ; *********** 8086 Flash LED program ***********
                        ; PROGRAM 20
                        ; flashes LED connected to bit 0 of Port B
                        ; ON for 0.5 s and OFF for 0.5 s continuously.
                        ; Uses circuit shown in Figure 6.8.
                        ; Program assumes clock frequency of 2 MHz.
                        ; Time for a 'T' state = 0.5 microseconds
                        ; ************************************************
                        ;
                        .MODEL      SMALL
                        .STACK
                        .DATA
                        ;
= 0010    ppi           EQU         10h             ; 8255 PPi base address
= 0011    leds          EQU         ppi+1           ; display (Port B)
= 0013    control       EQU         ppi+3           ; PPI control register
= 0080    iobyte        EQU         80h             ; I/O definition byte
= 0001    on            EQU         1               ; LED ON byte
= E5C6    h_sec         EQU         58822           ; 1/2 sec loop count
                        ;
                        .CODE
0400                    ORG         0400h           ; start of user RAM
                        ;
                        ; configure 8255 PPI
                        ;
0400  B0 80   start:    mov         al,iobyte       ; configure 8255 PPI
0402  E6 13             out         control,al      ; with Port B as outputs
                        ;
                        ; switch on LED
                        ;
0404  B0 01             mov         al,on           ; set bit 0 of Port B
0406  E6 11   flash:    out         leds,al         ; to switch on LED (10T)
                        ;
                        ; 0.5 second time delay (4T + {17T * 58821} + 5T)
                        ;
0408  B9 E5C6           mov         cx,h_sec        ; initialize delay count (5T)
040B  E2 FE   delay:    loop        delay           ; count down to zero (17T/5T)
                        ;
                        ; change state of LED
                        ;
040D  34 01             xor         al,1            ; invert bit 0      (3T)
040F  EB F5             jmp         flash           ; and repeat        (15T)
                        ;
                        END         start
```

8031

```
                        ; ********** 8031 Flash LED program **********
                        ; PROGRAM 21
                        ; flashes LED connected to bit 0 of Port B
                        ; ON for 0.5 s and OFF for 0.5 s continuously.
                        ; Uses circuit shown in Figure 6.8.
                        ; Program assumes clock frequency of 12 MHz.
                        ; Time for a 'T' state = 1.0 microseconds
                        ; **********************************************
                        ;
              ppi       EQU       0ff40h            ; 8255 PPi base address
              leds      EQU       ppi+1             ; display (Port B)
              control   EQU       ppi+3             ; PPi control register
              iobyte    EQU       80h               ; I/O definition byte
              on        EQU       1                 ; LED ON byte
              count1    EQU       40h               ; delay counter 1
              count2    EQU       41h               ; delay counter 2
              count3    EQU       42h               ; delay counter 3
              t_2µs     EQU       248               ; 2 µs count
              t_500µs   EQU       200               ; 100 milli-sec count
              t_100ms   EQU       5                 ; 500 milli-sec count
                        ;
8100                    ORG       8100h             ; start of user RAM
                        ;
                        ; configure 8255 PPI
                        ;
8100 90 FF 43  start:   mov       dptr,#control     ; point to PPI control register
8103 74 80              mov       a,#iobyte         ; configure 8255 PPI
8105 F0                 movx      @dptr,a           ; with Port B as outputs
                        ;
                        ; switch on LED
                        ;
8106 90 FF 41           mov       dptr,#leds        ; point to Port B
8109 74 01              mov       a,#on             ; set bit 0 of Port B
810B F0        flash:   movx      @dptr,a           ; to switch on LED
                        ;
                        ; 0.5 second time delay
                        ;
810C 75 40 05           mov       count1,#t_100ms   ; initialize dalay
810F 75 41 C8  dly1:    mov       count2,#t_500µs   ; counters 1, 2 and 3
8112 75 42 F8  dly2:    mov       count3,#t_2µs     ;
8115 D5 42 FD  dly3:    djnz      count3,dly3       ; count down 2 µs periods
8118 D5 41 F7           djnz      count2,dly2       ; count down 500 µs periods
811B D5 40 F1           djnz      count1,dly1       ; count down 100 ms periods
                        ;
                        ; change state of LED
                        ;
811E 65 01              xrl       a,1               ; invert bit 0
8120 21 0B              ajmp      flash             ; and repeat
                        ;
                        END       start
```

Table 6.2(a)   Z80

Instructions			Cycles	Times executed	Total cycles
h_sec = 41 665					
flash:	out	(leds),a	11 T	1	11 T
	ld	b,a	4 T	1	4 T
	ld	hl.h_sec	10 T	1	10 T
delay:	dec	hl	6 T	41 665	249 990 T
	ld	a,h	4 T	41 665	166 660 T
	or	I	4 T	41 665	166 660 T
	jp	nz,delay	10 T	41 665	416 650 T
	ld	a,b	4 T	1	4 T
	xor	1	7 T	1	7 T
	jp	flash	10 T	1	10 T
Total number of T states					1 000 006 T
Total execution time (1 000 006 x 0.5 µs)					500.003 ms

Table 6.2(b)   8086

Instructions		Cycles	Times executed	Total cycles
h_sec = 58 822				
flash:	out (leds),al	10 T	1	10 T
	mov cx,h_sec	4 T	1	4 T
delay:	loop delay	17 T	58 821	999 957 T
		5 T	1	5 T
	xor al	3 T	1	3 T
	jmp flash	15 T	1	15 T
Total number of T states				9999 994 T
Total execution time (999 994 x 0.5 µs)				499.997 ms

Table 6.2(c)   8031

Instructions		Cycles	Times executed	Total cycles
t_100 ms = 5 (outer loop)				
t_500 µs = 200 (middle loop)				
t_2 µs = 248 (inner loop)				
flash:	movx @dptr,a	2 T	1	2 T
	mov count1,#t_100 ms	2 T	1	2 T
dly1:	mov count2,#t_500 µs	2 T	1 x 5	10 T
dly2:	mov count3,#t_2 µs	2 T	1 x 200 x 5	2 000 T
dly3:	djnz count3,dly3	2 T	248 x 200 x 5	496 000 T
	djnz count2,dly2	2 T	200 x 5	2 000 T
	djnz count1,dly1	2 T	5	10 T
	xrl a,1	1 T	1	1 T
	ajmp flash	2 T	1	2 T

Total number of T states — 500 027 T
Total exeuction time (500 027 x 1.0 µs) — 500.007 ms

Activity

1   Devise an action list, structure chart, pseudo-code and program code for a program to cause the LEDs connected to Port B in Figure 6.8 to count up in **pure binary** at the rate of **one count per second**.

2   Construct a **trace table** for your program for the first one or two counts.

3   Using an appropriate microcomputer system, **test your program**, check for correct operation and hence verify the trace table.

Bit masking

You may have noticed that when reading port switches, all switches are read simultaneously and it is not possible to read just a single switch. This may be a problem since a program may require input in the form of:

1   a **single bit**, the state of which selects a particular action, e.g. motor ON if bit 0 = 1, motor OFF if bit 0 = 0,

2   a **selected group of bits**, e.g. if the range of inputs is from 0 to 7 then only three input bits are required and the remaining bits must be ignored, or

3   a **selection of different inputs**, all of which are valid. For example, if an upper or a lower case ASCII character 'a' is accepted, i.e. 'a' = 01100001, 'A' = 01000001, bit 5 may be either a 0 or a 1.

In all of these examples there are unused bits which must be ignored and these are put into a known state, usually logical 0. Individual bits may be changed to logical 0 by means of a **bit reset** instruction if this is included in the instruction set, but resetting selected groups of bits to logical 0 is a process known as **masking** and is carried out by means of the logical AND instruction. This instruction performs an AND operation between **corresponding bits** in its two operands, as shown in Figure 6.10.

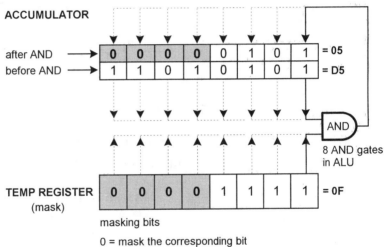

Figure 6.10

Example 7 – Bit masking

An interface circuit of the type shown in Figure 5.13 is connected to a microcomputer. A program is required to read input numbers 0 to 7 and show them as BCD on the 7-segment displays. If the input from S7 is a logical 1 the program terminates.

This program requires isolation of bit 7 to test for program termination, and masking of bits 4 to 7 for display purposes, but BCD correction is not required since the input never exceeds 7. A structure chart for this system is shown in Figure 6.11.

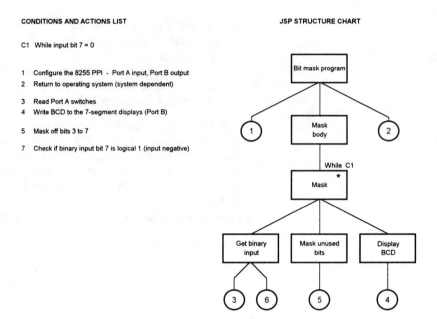

CONDITIONS AND ACTIONS LIST

C1   While input bit 7 = 0

1   Configure the 8255 PPI - Port A input, Port B output
2   Return to operating system (system dependent)

3   Read Port A switches
4   Write BCD to the 7-segment displays (Port B)

5   Mask off bits 3 to 7

7   Check if binary input bit 7 is logical 1 (input negative)

JSP STRUCTURE CHART

Figure 6.11

Pseudo-code for the bit masking program is as follows:

```
BIT MASK PROGRAM  seq
    DO  1                          {configure 8255 PPI}
    MASK BODY   iter  WHILE C1  {input not negative}
        MASK      seq
            DO  3                  {read switches}
            DO  6                  {test bit 7}
            DO  5                  {mask off bits 3 to 7}
            DO  4                  {write to 7-seg displays}
        MASK  end
    DO  2                          {return to monitor program}
    MASK BODY  end
BIT MASK PROGRAM   end
```

The following listings show typical assembly language programs for the Z80, 8086 and 8031 MPUs. Note how **bit 7** may be tested by checking for a **negative input**, since all negative numbers have their MSB = 1 in two's complement format.

Z80

```
                    ; *********** Z80 bit masking *********
                    ; PROGRAM 22
                    ; reads port switches and masks off bits
                    ; 3 to 7 and displays as BCD numbers 00
                    ; to 07 on the 7-segment LEDS.
                    ; Terminates if bit 7 is set.
                    ; Uses circuit shown in Figure 5.13
                    ; **************************************
                    ;
0010 =      ppi      EQU      10h          ; 8255 PPI base address
0010 =      switch   EQU      ppi          ; switches (Port A)
0011 =      leds     EQU      ppi+1        ; 7-seg displays (Port B)
0013 =      control  EQU      ppi+3        ; PPI control register
0090 =      iobyte   EQU      90h          ; I/O definition byte
0007 =      mask     EQU      00000111b    ; mask for bits 3 to 7
                    ;
1800               ORG      1800h         ; start of user RAM
                    ;
                    ; configure the 8255 PPI ports
                    ;
1800 3E 90  start:   ld       a, iobyte    ; set up Port A as inputs
1802 D3 13           out      (control),a  ; and Port B as outputs
                    ;
                    ; check for bit 7 set (negative input)
                    ;
1804 DB 10  read:    in       a,(switch)   ; read switches
1806 B7              or       a            ; set flags
1807 FA 11 18        jp       m,exit       ; if bit 7 = 1 then quit
                    ;
                    ; ignore switches S3 to S7
                    ;
180A E6 07           and      mask         ; bits 3 to 7 now all 0
                    ;
                    ; show input number on the 7-segment displays
                    ;
180C D3 11           out      (leds),a     ; show result
180E C3 04 18        jp       read         ; and repeat
                    ;
                    ; exit if negative input applied
                    ;
1811 F7     exit:    rst      30h          ; return to minitor program
                    ;
                    END      start
```

8086

```
                          ; ************ 8086 bit masking **********
                          ; PROGRAM 23
                          ; reads port switches and masks off bits
                          ; 3 to 7 and displays as BCD numbers 00
                          ; to 07 on the 7-segment LEDS.
                          ; Terminates if bit 7 is set.
                          ; Uses circuit shown in Figure 5.13
                          ; ****************************************
                          ;
                          .MODEL     SMALL
                          .STACK
                          ;
= 0010        ppi         EQU        10h          ; 8255 PPI base address
=             switch      EQU        ppi          ; switches (Port A)
= 0011        leds        EQU        ppi+1        ; 7-seg displays (Port B)
= 0013        control     EQU        ppi+3        ; PPI control register
= 0090        iobyte      EQU        90h          ; I/O definition byte
= 0007        mask1       EQU        00000111b    ; mask for bits 3 to 7
                          ;
                          .CODE
0400                      ORG        0400h        ; start of user RAM
                          ;
                          ; configure the 8255 PPI ports
                          ;
0400 B0 90    start:      mov        al,iobyte    ; set up Port A as inputs
0402 E6 13                out        control,al   ; and Port B as outputs
                          ;
                          ; check for bit 7 set (negative input)
                          ;
0404 E4 10    read:       in         al,switch    ; read switches
0406 0A C0                or         al,al        ; set flags
0408 78 06                js         exit         ; if bit 7 = 1 then quit
                          ;
                          ; ignore swtiches S3 to S7
                          ;
040A 24 07                and        al,mask1     ; bits 3 to 7 now all 0
                          ;
                          ; show input number on the 7-segment displays
                          ;
040C E6 11                out        leds,al      ; show result
040E EB F4                jmp        read         ; and repeat
                          ;
                          ; exit if negative input applied
                          ;
0410 CD 20    exit:       int        20h          ; return to monitor program
                          ;
                          END        start
```

8031

```
                           ; *********** 8031 bit masking **********
                           ; PROGRAM 24
                           ; reads port switches and masks off bits
                           ; 3 to 7 and displays as BCD numbers 00
                           ; to 07 on the 7-segment LEDS.
                           ; Terminates if bit 7 is set.
                           ; Uses circuit shown in Figure 5.13
                           ; ****************************************
                           ;
              ppi          EQU        0ff40h          ; 8255 PPI base address
              switch       EQU        ppi             ; switches (Port A)
              leds         EQU        ppi+1           ; 7-seg displays (Port B)
              control      EQU        ppi+3           ; PPI control register
              iobyte       EQU        90h             ; I/O definition byte
              mask1        EQU        00000111b       ; mask for bits 3 to 7
              acc.7        EQU        0e7h            ; bit 7 of accumulator
              monitor      EQU        0a2h            ; warm restart
                           ;
8100                       ORG        8100h           ; start of user RAM
                           ;
                           ; configure the 8255 PPI ports
                           ;
8100 90 FF 43  start:      mov        dptr,#control   ; point to PPI control reg
8103 74 90                 mov        a,#iobyte       ; set up Port A as inputs
8105 F0                    movx       @dptr,a         ; and Port B as outputs
                           ;
                           ; check for bit 7 set (negative input)
                           ;
8106 90 FF 40  read:       mov        dptr,#switch    ; point to switches
8109 E0                    movx       a,@dptr         ; read switches
810A 20 E7 06              jb         acc.7,exit      ; if acc bit 7 = 1 then quit
                           ;
                           ; ignore switches S3 to S7
                           ;
810D 55 07                 anl        a,mask1         ; bits 3 to 7 now all 0
                           ;
                           ; show input number on the 7-segment displays
                           ;
810F A3                    inc        dptr            ; point to LEDs
8110 F0                    movx       @dptr,a         ; show result
8111 80 F3                 sjmp       read            ; and repeat
                           ;
                           ; exit if negative input applied
                           ;
8113 02 00 A2  exit:       ljmp       monitor         ; return to monitor program
                           ;
                           END        start
```

Activity

1   Connect an interface circuit of the type shown in Figure 6.8 to a microcomputer.

2   A unit is required to provide **time delays** between 0 and 30 seconds, using a binary counter that decrements to zero at one second intervals from an initial count

entered via the binary switch inputs. Any initial setting greater than 30 must be ignored.

3 Devise an **action list**, **pseudo-code** and **program code** to enable the system to operate as described.

4 Using an appropriate microcomputer system, **test your program**, and check for correct operation.

Waiting for inputs to change

A microcomputer may control an external system by changing logic levels applied to the system, but the effect of such changes may or may not be important to the microcomputer.

A microcomputer may, for example, switch on a heating system for 10 minutes but be unconcerned about the actual temperature attained, or the system itself could have built-in temperature control. In either case, the microcomputer is concerned solely with switching a system on or off.

However, if a microcomputer is to regulate temperature by switching a heating element on and off as necessary, then **feedback** is necessary to provide the microcomputer with information about temperature within the system. In such a system the microcomputer would:

1 switch on the heating element then wait for an input to change, indicating upper temperature limit reached,

2 switch off the heater, then wait for an input to change, indicating lower temperature limit reached,

3 repeat 1 and 2 continuously.

Example 8 – Waiting for inputs to change

The simple temperature control interface shown in Figure 6.12 is used to maintain temperature in an industrial control system at a fixed level. The thermostat contacts **open** when temperature reaches an **upper limit**, and **close** when it drops below a **lower limit**. A program is required to operate the heater so that average temperature is maintained at the correct level.

A structure chart for this system is shown in Figure 6.13. It should be noted that this structure is intended to demonstrate the principle of waiting for inputs to change and does not represent the most efficient way of solving this

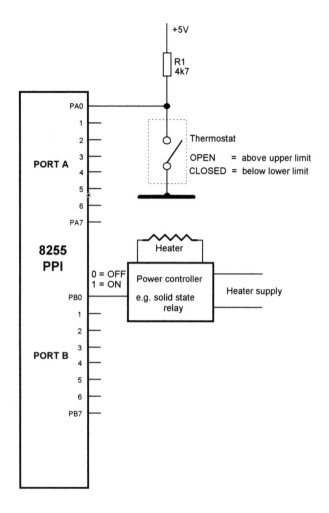

Figure 6.12

problem. In particular, duplication in actions 2 and 3 could be prevented by restructuring the chart!

Pseudo-code for the temperature control program is as follows:

```
TEMPERATURE CONTROL PROGRAM   seq
    DO  1                           {configure 8255 PPI}
    CONTROLLER BODY iter  WHILE  C1  {always}
        CONTROLLER    seq
            WAIT FOR CLOSURE BODY iter  WHILE C2 {open}
                WAIT FOR THERMOSTAT TO CLOSE    seq
                    DO  2               {read thermostat}
                    DO  6               {check status}
                WAIT FOR THERMOSTAT TO CLOSE    end
            WAIT FOR CLOSURE BODY   end
            DO   4                          {switch heater on}
```

```
        WAIT FOR OPENING BODY  iter  WHILE C3{closed}
            WAIT FOR THERMOSTAT TO OPEN    seq
                 DO  3                {read thermostat}
                 DO  7                {check status}
            WAIT FOR THERMOSTAT TO OPEN    end
        WAIT FOR OPENING BODY  end
        DO    5                         {switch heater off}
      CONTROLLER  end
    CONTROLLER BODY  end
  TEMPERATURE CONTROL PROGRAM   end
```

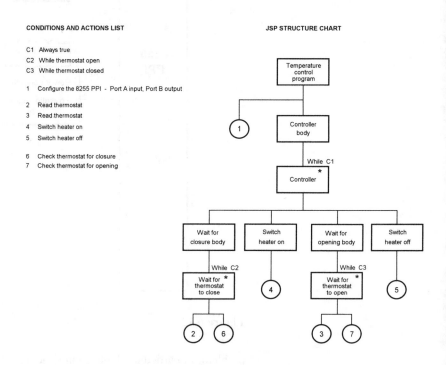

CONDITIONS AND ACTIONS LIST

C1   Always true
C2   While thermostat open
C3   While thermostat closed

1   Configure the 8255 PPI - Port A input, Port B output

2   Read thermostat
3   Read thermostat
4   Switch heater on
5   Switch heater off

6   Check thermostat for closure
7   Check thermostat for opening

Figure 6.13

The following listings show typical assembly language programs for the Z80, 8086 and 8031 MPUs.

Z80

```
                        ; ****** Z80 wait for input change ******
                        ; PROGRAM 25
                        ; controls heating element attached to
                        ; Port B bit 2.
                        ; Waits for input change from thermostat
                        ; attached to bit 3 of Port A.
                        ; Uses circuit shown in Figure 6.12
                        ; ***************************************
                        ;
0010 =      ppi         EQU     10h             ; 8255 PPI base address
0010 =      stat        EQU     ppi             ; switches (Port A)
0011 =      heater      EQU     ppi+1           ; 7-seg displays (Port B)
0013 =      control     EQU     ppi+3           ; PPI control register
0090 =      iobyte      EQU     90h             ; I/O definition byte
0002 =      on          EQU     00000111b       ; ON state for heater
0000 =      off         EQU     0               ; OFF state for heater
                        ;
1800                    ORG     1800h           ; start of user RAM
                        ;
                        ; configure the 8255 PPI ports
                        ;
1800 3E 90  start:      ld      a,iobyte        ; set up Port A as inputs
1802 D3 13              out     (control),a     ; and Port B as outputs
                        ;
                        ; wait for thermostat to close
                        ;
1804 DB 10  check1:     in      a,(stat)        ; read thermostat
1806 CB 5F              bit     3,a             ; test its condition
1808 C2 04 18           jp      nz,check1       ; thermostat close?
                        ;
                        ; switch on heater
                        ;
180B 3E 02              ld      a,on            ; set heater bit
180D D3 11              out     (heater),a      ; to switch it on
                        ;
                        ; wait for thermostat to open
                        ;
180F DB 10  check2:     in      a,(stat)        ; read thermostat
1811 CB 5F              bit     3,a             ; test its condition
1813 CA 0F 18           jp      z,check2        ; thermostat close?
                        ;
                        ; switch off heater
                        ;
1816 3E 00              ld      a,off           ; reset heater bit
1818 D3 11              out     (heater),a      ; to switch it off
181A C3 04 18           jp      check1          ; continuous control
                        ;
                        END     start
```

8086

```
                         ; ****** 8086 wait for input change ******
                         ; PROGRAM 26
                         ; controls heating element attached to
                         ; Port B bit 2.
                         ; Waits for input change from thermostat
                         ; attached to bit 3 of Port A.
                         ; Uses circuit shown in Figure 6.12
                         ; *****************************************
                         ;
                         .MODEL        SMALL
                         .STACK
                         ;
=     0010      ppi      EQU           10h            ; 8255 PPI base address
=               stat     EQU           ppi            ; switches (Port A)
=     0011      heater   EQU           ppi+1          ; 7-seg displays (Port B)
=     0013      control  EQU           ppi+3          ; PPI control register
=     0090      iobyte   EQU           90h            ; I/O definition byte
=     0002      on       EQU           00000010b      ; ON state for heater
=     0000      off      EQU           0              ; OFF state for heater
                         ;
                         .CODE
0400                     ORG           0400h          ; start of user RAM
                         ;
                         ; configure the 8255 PPI ports
                         ;
0400 B0 90     start:    mov           al,iobyte      ; set up Port A as inputs
0402 E6 13               out           control,al     ; and Port B as outputs
                         ;
                         ; wait for thermostat to close
                         ;
0404 E4 10     check1:   in            al,stat        ; read thermostat
0406 A8 08               test          al,8           ; test its condition
0408 75 FA               jnz           check1         ; thermostat close?
                         ;
                         ; switch on heater
                         ;
040A B0 02               mov           al,on          ; set heater bit
040C E6 11               out           heater,al      ; to switch it on
                         ;
                         ; wait for thermostat to open
                         ;
040E E4 10     check2:   in            al,stat        ; read thermostat
0410 A8 08               test          al,8           ; test its condition
0412 74 FA               jz            check2         ; thermostat close?
                         ;
                         ; switch off heater
                         ;
0414 B0 00               mov           al,off         ; reset heater bit
0416 E6 11               out           heater,al      ; to switch it off
0418 EB EA               jmp           check1         ; continuous control
                         ;
                         END           start
```

8031

```
                          ; ****** 8031 wait for input change ******
                          ; PROGRAM 27
                          ; controls heating element attached to
                          ; Port B bit 2.
                          ; Waits for input change from thermostat
                          ; attached to bit 3 of Port A.
                          ; Uses circuit shown in Figure 6.12
                          ; ****************************************
                          ;
              ppi         EQU         0ff40h          ; 8255 PPI base address
              stat        EQU         ppi             ; switches (Port A)
              control     EQU         ppi+3           ; PPI control register
              iobyte      EQU         90h             ; I/O definition byte
              acc.3       EQU         0e3h            ; bit 3 of accumulator
              on          EQU         00000010b       ; ON state for heater
              off         EQU         0               ; OFF state for heater
                          ;
8100                      ORG         8100h           ; start of user RAM
                          ;
                          ; configure the 8255 PPI ports
                          ;
8100 90 FF 43 start:      mov         dptr,#control   ; point to PPI control reg
8103 74 90                mov         a,#iobyte       ; set up Port A as inputs
8105 F0                   movx        @dptr,a         ; and Port B as outputs
                          ;
                          ; wait for thermostat to close
                          ;
8106 90 FF 40 check1:     mov         dptr,#stat      ; point to thermostat
8109 E0                   movx        a,@dptr         ; read thermostat
                          ;
810A 20 E3 F9             ; switch on heater
                          ;
810D A3                   inc         dptr            ; point to heater port
810E 74 02                mov         a,#on           ; set heater bit
8110 F0                   movx        @dptr,a         ; to switch it on
                          ;
                          ; wait for thermostat to open
                          ;
8111 90 FF 40 check2:     mov         dptr,#stat      ; point to thermostat
8114 E0                   movx        a,@dptr         ; read thermostat
8115 30 E3 F9             jnb         acc.3,check2    ; thermostat close?
                          ;
                          ; switch off heater
                          ;
8118 A3                   inc         dptr            ; point to heater port
8119 74 00                mov         a,#off          ; reset heater bit
811B F0                   movx        @dptr,a         ; to switch it off
811C 80 E8                sjmp        check1          ; continuous control
                          ;
                          END         start
```

Pulse counting

Very often the timing of a program must be controlled by randomly occurring external events, for example counting items on a conveyor belt or counting revolutions of a motor shaft. In such applications a sensor is used to monitor passing objects, and the count must be increased by one for each object passing the sensor. If the system were based upon a change in input level only, then counting would continue all the time that the sensor detected the object, i.e. each object would be counted many times. A single object passing the sensor causes its output to change in logic level and then change back again, i.e. a *pulse* is generated. When a microcomputer is used for counting objects it must therefore **count the pulses** generated by the sensor and this entails monitoring two opposite direction changes in logic level for each pulse (see Figure 6.14).

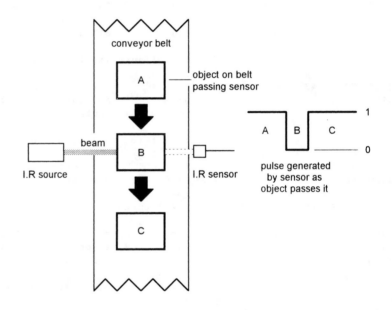

Figure 6.14

Example 9 – Pulse counting

This example uses the interface circuit shown in Figure 5.13.
   A program is required to make the system operate as a **BCD counter** which starts at zero and is incremented by one each time that S0 is operated (any digital sensing device could be used in place of S0). The program is terminated when the count reaches 100.

A structure chart for this system is shown in Figure 6.15.

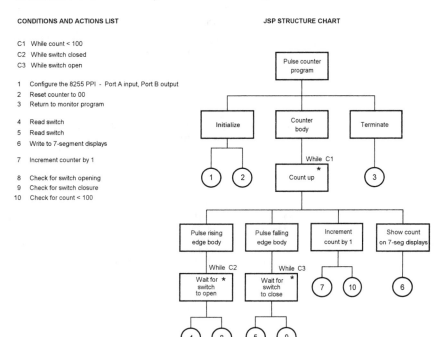

CONDITIONS AND ACTIONS LIST

C1 While count < 100
C2 While switch closed
C3 While switch open

1   Configure the 8255 PPI - Port A input, Port B output
2   Reset counter to 00
3   Return to monitor program

4   Read switch
5   Read switch
6   Write to 7-segment displays

7   Increment counter by 1

8   Check for switch opening
9   Check for switch closure
10 Check for count < 100

Figure 6.15

Pseudo-code for the pulse counting program is as follows:

```
PULSE COUNTER PROGRAM   seq
    DO  1                        {configure 8255 PPI}
    DO  2                        {reset counter to 00}
    COUNTER BODY  iter  WHILE C1       {count < 100}
        COUNT UP    seq
            PULSE RISING EDGE BODY iter WHILE C2 {S0 closed}
                DO  4        {read switch}
                DO  8        {check for S0 opening}
            PULSE RISING EDGE BODY  end
            PULSE FALLING EDGE BODY iter WHILE C3 {S0 open}
                DO  5        {read switch}
                DO  9        {check if S0 closed}
            PULSE FALLING EDGE BODY  end
            DO       7            {BCD increment count by 1}
            DO       10           {check for count < 100}
            DO       6            {show count on 7-seg disp}
        COUNT UP    end
    COUNTER BODY  end
    DO  3                            {return to monitor program}
PULSE COUNTER PROGRAM   end
```

The following listings show typical assembly language programs for the Z80, 8086 and 8031 MPUs. Note that some switches '**bounce**', and each operation of the switch results in a sequence of pulses being generated. If this problem occurs then a **short delay (10–20 ms)** between switch readings may be necessary.

Z80

```
                        ;  ********** Z80 pulse counting ***********
                        ;  PROGRAM 28
                        ;  counts pulses applied to Port A bit 0.
                        ;  Displays BCD count 00 to 99 on two
                        ;  7-segment displays attached to Port B.
                        ;  Uses circuit shown in Figure 5.13
                        ;  ****************************************
                        ;
0010 =        ppi        EQU        10h                ; 8255 PPI base address
0010 =        switch     EQU        ppi                ; switches (Port A)
0011 =        display    EQU        ppi+1              ; 7-seg displays (Port B)
0013 =        control    EQU        ppi+3              ; PPI control register
0090 =        iobyte     EQU        90h                ; I/O definition byte
                        ;
1800                     ORG        8100h              ; start of user RAM
                        ;
                        ; initialize system
                        ;
1800 3E 90    start:     ld         a,iobyte           ; set up Port A as inputs
1802 D3 13               out        (control),a        ; and Port B as outputs
1804 0E 00               ld         c,0                ; zero counter
                        ;
                        ; show count on 7-segment displays
                        ;
1806 79                  ld         a,c                ; show count
1807 D3 11    count:     out        (display),a        ; on 7-seg displays
                        ;
                        ; wait for switch S0 to open
                        ;
1809 DB 10    check1:    in         a,(switch)         ; read the switches
180B CB 47               bit        0,a                ; test S0 state
180D CA 09 18            jp         z,check1           ; switch still closed?
                        ;
                        ; wait for switch S0 to close
                        ;
1810 DB 10    check2:    in         a,(switch)         ; read the switches
1812 CB 47               bit        0,a                ; test S0 state
1814 C2 10 18            jp         nz,check2          ; switch still open?
                        ;
                        ; BCD increment of counter
                        ;
1817 79                  ld         a,c                ; get current count
1818 C6 01               add        a,1                ; increment count by 1
181A 27                  daa                           ; in BCD form
181B 4F                  ld         c,a                ; and updata counter
                        ;
                        ; repeat if < 100
                        ;
181C D2 07 18            jp         nc,count           ; exit if > 99
                        ;
181F F7       exit:      rst        30h                ; return to monitor
                        ;
                         END        start
```

8086

```
                          ; ********* 8086 pulse counting *********
                          ; PROGRAM 29
                          ; counts pulses applied to Port A bit 0.
                          ; Displays BCD count 00 to 99 on two
                          ; 7-segment displays attached to Port B.
                          ; Uses circuit shown in Figure 5.13
                          ; *****************************************
                          ;
                          .MODEL      SMALL
                          .STACK
                          ;
=    0010       ppi       EQU         10h             ; 8255 PPI base address
=               switch    EQU         ppi             ; switches (Port A)
=    0011       display   EQU         ppi+1           ; 7-seg displays (Port B)
=    0013       control   EQU         ppi+3           ; PPI control register
=    0090       iobyte    EQU         90h             ; I/O definition byte
                          ;
                          .CODE
0400                      ORG         0400h           ; start of user RAM
                          ;
                          ; initialize system
                          ;
0400 B0 90     start:     mov         al,iobyte       ; set up Port A as inputs
0402 E6 13                out         control, al     ; and Port B as outputs
0404 B4 00                mov         ah,0            ; zero count
                          ;
                          ; show count on 7-segment displays
                          ;
0406 8A C4     count:     out         display,al      ; on 7-segment displays
                          ;
                          ; wait for switch S0 to open
                          ;
040A E4 10     check1:    in          al,switch       ; read the switches
040C A8 01                test        al,1            ; test S0 state
040E 74 FA                jz          check1          ; switch still closed?
                          ;
                          ; wait for switch S0 to close
                          ;
0410 E4 10     check2:    in          al,switch       ; read the switches
0412 A8 01                test        al,1            ; test S0 state
0414 75 FA                jnz         check2          ; switch still open?
                          ;
                          ; BCD increment of counter
                          ;
0416 8A C4                mov         al,ah           ; get current count
0418 04 01                add         al,1            ; increment count by 1
041A 27                   daa                         ; in BCD form
041B 8A E0                mov         ah,al           ; and updata counter
                          ;
                          ; repeat if < 100
                          ;
041D 73 E9                jnc         count           ; exit if > 99
                          ;
041F CD 20     exit:      int         20h             ; return to monitor
                          ;
                          END         start
```

8031

```
                              ; ********** 8031 pulse counting **********
                              ; PROGRAM 30
                              ; counts pulses applied to Port A bit 0.
                              ; Displays BCD count 00 to 99 on two
                              ; 7-segment displays attached to Port B.
                              ; Uses circuit shown in Figure 5.13
                              ; ****************************************
                              ;
                 ppi          EQU       0ff40h            ; 8255 PPI base address
                 switch       EQU       ppi               ; switches (Port A)
                 display      EQU       ppi+1             ; 7-seg displays (Port B)
                 control      EQU       ppi+3             ; PPI control register
                 iobyte       EQU       90h               ; I/O definition byte
                 acc.0        EQU       0e0h              ; bit 0 of accumulator
                 b            EQU       0f0h              ; auxiliary accumulator
                 monitor      EQU       0a2h              ; warm restart
                              ;
8100                          ORG       8100h             ; start of user RAM
                              ;
                              ; initialize system
                              ;
8100  90 FF 43   start:       mov       dptr,#control     ; pointer to PPI control reg
8103  74 90                   mov       a,#iobyte         ; set up Port A as inputs
8105  F0                      movx      @dptr,a           ; and Port B as outputs
8106  75 F0 00                mov       b,#0              ; zero count
                              ;
                              ; show count on 7-segment displays
                              ;
8109  90 FF 41                mov       dptr,#display     ; point to 7-seg displays
810C  E5 F0                   mov       a,b               ; copy count to acc
810E  F0         count:       movx      @dptr,a           ; and show it
                              ;
                              ; wait for switch S0 to open
                              ;
810F  90 FF 40                mov       dptr,#switch      ; point to switch
8112  E0         check1:      movx      a,@dptr           ; read the switches
8113  30 E0 FC                jnb       acc.0,check1      ; switch still closed?
                              ;
                              ; wait for switch S0 to close
                              ;
8116  E0         check2:      movx      a,@dptr           ; read the switches
8117  20 E0 FC                jb        acc.0,check2      ; switch still open?
                              ;
                              ; BCD increment of counter
                              ;
811A  E5 F0                   mov       a,b               ; get current count
811C  25 01                   add       al,1              ; increment count by 1
811E  D4                      da        a                 ; in BCD form
811F  F5 F0                   mov       b,a               ; and update counter
                              ;
                              ; repeat if < 100
                              ;
8121  A3                      inc       dptr              ; point to displays
8122  50 EA                   jnc       count             ; exit if > 99
                              ;
8124  02 00 A2   exit:        ljmp      monitor           ; return to monitor
                              ;
                              END       start
```

Activity

1   Using an interface circuit of the type shown in Figure
6.8, connect the TTL output of a square wave
generator to Port A bit 0 (make sure that S0 is left in the
open position).

2   A program is required to cause a **16-bit down counter**
to be decremented by each input pulse from the signal
generator. The counter must first be preset with 600_{10},
6000_{10} or 60000_{10} so that input signals of 10 Hz,
100 Hz or 1 kHz should all take 1 minute to decrement
the counter to zero. The LED connected to bit 0 of Port
B is used to indicate start and finish of the count, thus
enabling the frequency of the signal generator to be
checked.

3   Devise an **action list**, **structure chart**, **pseudo-code**
and **program code** to enable the system to operate as
described.

4   Using an appropriate microcomputer system, **test
your program**, and check for correct operation.

Sequencing

The processes investigated so far have all been of a type that could be dealt
with by the application of simple mathematical principles, i.e. there was a
direct relationship between inputs and outputs. In many applications,
outputs must change in a particular sequence, but each step in the sequence
has **no logical or mathematical relationship** to the previous step. In such
cases the only practical solution is to store the sequence of bit patterns in
memory, and transfer these one after another as required by the system. This
is a process known as **sequencing**, and the stored bit patterns are known as
a **look-up table**. Such a table may also be randomly accessed for code
translations such as BCD to 7-segment codes.

Example 10 – Sequencing

The interface circuit shown in Figure 6.16 represents the traffic
lights at a road junction. A program is required to control the
lights according to the sequence and relative timings shown in
Table 6.3. This table clearly shows the lack of any obvious
relationship between successive steps in the sequence.

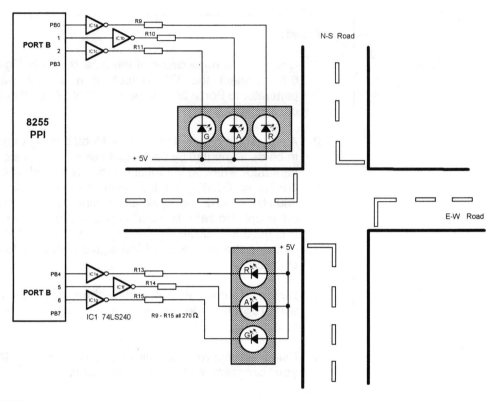

Figure 6.16

Table 6.3

STEP	E–W	7	6	5	4	N–S	3	2	1	0	TIME
		–	G	A	R		–	G	A	R	
1	G	X	1	0	0	R	X	0	0	1	10
2	A	X	0	1	0	R	X	0	0	1	2
3	R	X	0	0	1	R	X	0	0	1	1
4	R	X	0	0	1	RA	X	0	1	1	2
5	R	X	0	0	1	G	X	1	0	0	10
6	R	X	0	0	1	A	X	0	1	0	2
7	R	X	0	0	1	R	X	0	0	1	1
8	RA	X	0	1	1	R	X	0	0	1	2
9	G	X	1	0	0	R	X	0	0	1	10 etc

X = Don't care

The timings in this example represent seconds and have therefore been deliberately shortened to speed up the testing process. In practice the times would be much longer, and may also be variable, changing throughout the day according to traffic demands. A conditions and actions list for the lights, together with the structure chart, is shown in Figure 6.17.

The structure for the time delay has not been shown in full but will be similar to that shown in Figure 6.7.

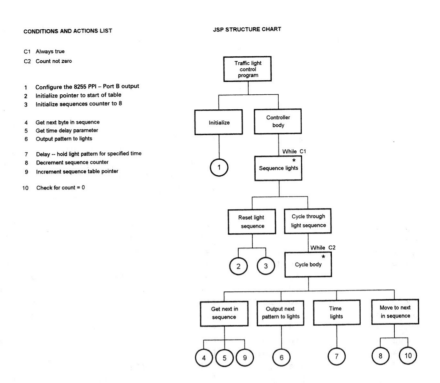

Figure 6.17

The pseudo-code for the traffic light program is as follows:

```
TRAFFIC LIGHT CONTROL PROGRAM  seq
    DO  1                          {configure 8255 PPI}
    CONTROLLER BODY  iter  WHILE C1 {repeat continuously}
        SEQUENCE LIGHTS  seq
            DO  2                  {point to start of table}
            DO  3                  {initialize counter to 8}
            CYCLE iter  WHILE C2      {count <> 0}
                CYCLE BODY  seq
                    DO  4   {get next pattern}
                    DO  5   {get next delay parameter}
                    DO  6   {output pattern to lights}
                    DO  9   {increment sequence pointer}
                    DO  7   {timing delay for lights}
                    DO  8   {decrement seq. counter}
                    DO  10  {check for count = 0}
                CYCLE BODY  end
            CYCLE  end
        SEQUENCE LIGHTS  end
    CONTROLLER BODY  end
TRAFFIC LIGHT CONTROL PROGRAM  end
```

The following listings show typical assembly language programs for the Z80, 8086 and 8031 MPUs.

Z80

```
                    ; ************ Z80 Sequencing *************
                    ; PROGRAM 31
                    ; Traffic light controller program for
                    ; two sets of lights attached to Port B.
                    ; Assumes 2 MHz clock.
                    ; Uses circuit shown in Figure 6.16
                    ; *****************************************
                    ;
0010 =        ppi       EQU     10h            ; 8255 PPI base address
0011 =        lights    EQU     ppi+1          ; traffic lights
0013 =        control   EQU     ppi+3          ; PPI control register
0090 =        iobyte    EQU     90h            ; I/O definition byte
61A8 =        hsec      EQU     25000          ; 0.5 s delay parameter
0010 =        red_ew    EQU     10h            ; E-W red
0020 =        amber_ew  EQU     20h            ; E-W amber
0040 =        green_ew  EQU     40h            ; E-W green
0001 =        red_ns    EQU     1              ; N-S red
0002 =        amber_ns  EQU     2              ; N-S amber
0004 =        green_ns  EQU     4              ; N-S green
0002 =        one_sec   EQU     2*1            ; 1 second
0004 =        two_sec   EQU     one_sec*2      ; 2 seconds
0014 =        ten_sec   EQU     one_sec*10     ; 10 seconds
                    ;
1800                    ORG     1800h          ; start of user RAM
                    ;
                    ; initialize system
                    ;
1800 3E 90    start:    ld      a,iobyte       ; set up Port A as inputs
1802 D3 13              out     (control),a    ; and Port B as outputs
                    ;
                    ; reset light sequencer
                    ;
1804 DD 21 30 18 cycle: ld      ix,step        ; pointer to lights table
1808 06 08              ld      b,8            ; initialize counter
                    ;
                    ; display next pattern in sequence
                    ;
180A DD 7E 00  next:    ld      a,(ix+0)       ; get display pattern
180D D3 11              out     (lights),a     ; output light pattern
                    ;
                    ; hold pattern for set time period
                    ;
180F DD 4E 01           ld      c,(ix+1)       ; get delay parameter
1812 CD 1E 18           call    delay          ; wait for selected time
                    ;
                    ; move pointer and check for end of table
                    ;
1815 DD 23              inc     ix             ; point to next pair
1817 DD 23              inc     ix             ; in sequence
1819 10 EF              djnz    next           ; and repeat, if end of table
181B C3 04 18           jp      cycle          ; start again
```

```
                        ;
                        ; time delay, number 0.5 secs passed via reg C
                        ;
                        ld      hl,hsec     ; 0.5 second delay
1821  2B      dly:      dec     hl          ; count down to zero
1822  ED 44             neg                 ; does nothing except add
1824  ED 44             neg                 ; 16 T states to loop
1826  7C               ld      a,h          ; set Z flag if hl=0
1827  B5               or      l            ; dec hl sets no flags
1828  C2 21 18         jp      nz,dly       ; repeat until hl=0
182B  0D               dec     c            ; count down 0.5 sec
182C  C2 1E 18         jp      nz,delay     ; periods
182F  C9               ret                  ; go back to main program
                        ;
                        ; traffic light patterns and delay times
                        ;
1803  41 14   step:     DEFB    green_ew + red_ns, ten_sec
1832  21 04             DEFB    amber_ew + red_ns, two_sec
1834  11 02             DEFB    red_ew   + red_ns, one_sec
1836  13 04             DEFB    red_ew   + amber_ns + red_ns,two_sec
1838  14 14             DEFB    red_ew   + green_ns, ten_sec
183A  12 04             DEFB    red_ew   + amber_ns, two_sec
183C  11 02             DEFB    red_ew   + red_ns, one_sec
183E  31 04             DEFB    amber_ew + red_ew + red_ns,two_sec
                        ;
                        END     start
```

8086

```
                        ; *********** 8086 Sequencing ***********
                        ; PROGRAM 32
                        ; Traffic light controller program for
                        ; two sets of lights attached to Port B.
                        ; Assumes 2 MHz clock.
                        ; Uses circuit shown in Figure 6.16
                        ; ****************************************
                        ;
                        .MODEL   SMALL
                        .STACK
                        ;
=    0010   ppi         EQU      10h             ; 8255 PPI base address
=    0011   lights      EQU      ppi+1           ; traffic lights
=    0013   control     EQU      ppi+3           ; PPI control register
=    0090   iobyte      EQU      90h             ; I/O definition byte
=    C350   hsec        EQU      50000           ; 0.5 s delay parameter
=    0010   red_ew      EQU      10h             ; E-W red
=    0020   amber_ew    EQU      20h             ; E-W amber
=    0040   green_ew    EQU      1               ; N-S red
=    0002   amber_ns    EQU      2               ; N-S amber
=    0004   green_ns    EQU      4               ; N-S green
=    0014   ten_sec     EQU      10*2            ; 10 seconds
=    0004   two_sec     EQU      2*2             ;  2 seconds
=    0002   one_sec     EQU      1*2             ;  1 second
                        ;
                        .CODE
0400                    ORG      0400h           ; start of user RAM
```

```
                          ;
                          ; initialize system
                          ;
0400  B8 ---- R   start:  mov     ax,@DATA       ; set up data segment
0403  8E D8               mov     ds,ax          ;
0405  B0 90               mov     al,iobyte      ; set up Port A as inputs
0407  E6 13               out     control,al     ; and Port B as outputs
                          ;
                          ; reset light sequencer
                          ;
0409  BB 0000 R   cycle:  mov     bx,OFFSET step ; pointer to lights table
040C  B9 0008             mov     cx,8           ; initialize counter
                          ;
                          ; display next pattern in sequence
                          ;
040F  8B 07       next:   mov     ax,[bx]        ; al = display pattern
0411  E6 11               out     lights,al      ; ah = time parameter
0413  E8 041C R           call    delay          ; wait for selected time
                          ;
                          ; move pointer and check for end of table
                          ;
0416  43                  inc     bx             ; point to next pair
0417  43                  inc     bx             ; in sequence
0418  E2 F5               loop    next           ; and repeat, if end of table
041A  EB ED               jmp     cycle          ; start again
                          ;
                          ; time delay, number 0.5 secs passed via reg C
                          ;
041C              delay:  PROC    NEAR
041C  BA C350             mov     dx,hsec        ; 0.5 second delay
041F  4A          dly:    dec     dx             ; count down to zero
0420  8A E4               mov     ah,ah          ; add 2 T states to loop
0422  75 FB               jnz     dly            ; repeat until dx=0
0424  FE CC               dec     ah             ; count down 0.5 sec
0426  75 F4               jnz     delay          ; periods
0428  C3                  ret                    ; go back to main program
0429              delay:  ENDP
                          ;
                          .DATA
                          ;
                          ; traffic light patterns and delay times
                          ;
0000  41 14       step:   DB      green_ew + red_ns, ten_sec
0002  21 04               DB      amber_ew + red_ns, two_sec
0004  11 02               DB      red_ew      + red_ns, one_sec
0006  13 04               DB      red_ew      + amber_ns    + red_ns,two_
sec
0008  14 14               DB      red_ew      + green_ns, ten_sec
000A  12 04               DB      red_ew      + amber_ns, two_sec
000C  11 02               DB      red_ew      + red_ns, one_sec
000E  31 04               DB      amber_ew    + red_ew      + red_ns,two_
sec
                          ;
                          END     start
```

8031

```
                          ; *********** 8031 Sequencing ***********
                          ; PROGRAM 33
                          ; Traffic light controller program for
                          ; two sets of lights attached to Port B.
                          ; Assumes 12 MHz clock.
                          ; Uses circuit shown in Figure 6.16
                          ; ****************************************
                          ;
                   ppi      EQU    0ff40h              ; 8255 PPI base address
                   lights   EQU    ppi+1               ; traffic lights
                   control  EQU    ppi+3               ; PPI control register
                   red_ew   EQU    10h                 ; E-W red
                   amber_ew EQU    20h                 ; E-W amber
                   green_ew EQU    40h                 ; E-W green
                   red_ns   EQU    1                   ; N-S red
                   amber_ns EQU    2                   ; N-S amber
                   green_ns EQU    4                   ; N-S green
                   t_2µs    EQU    248                 ; 2 µs period count
                   t_500µs  EQU    200                 ; 500 µs period count
                   one_sec  EQU    10                  ; 1 second
                   two_sec  EQU    20                  ; 2 seconds
                   ten_sec  EQU    100                 ; 10 seconds
                          ;
8100                      ORG    8100h                ; start of user RAM
                          ;
                          ; initialize system
                          ;
8100  90 FF 43   start:   mov    dptr,#control        ; set up Port B as outputs
8103  74 90               mov    a,#iobyte            ;
8105  F0                  movx   @dptr,a              ; and Port B as outputs
8106  90 FF 41            mov    dptr,#lights         ; point to lights
                          ;
                          ; reset light sequencer
                          ;
8109  78 43     cycle:    mov    r0,#step             ; pointer to lights table
810B  79 4B               mov    r1,#delays           ; pointer to delays table
810D  7A 08               mov    r2,#8                ; initialize counter
                          ;
                          ; display next pattern in sequence
                          ;
810F  E6        next:     mov    a,@r0                ; get display pattern
8110  F0                  movx   @dptr,a              ; output light pattern
                          ;
                          ; hold pattern for set time period
                          ;
8111  87 40               mov    count1,@r1           ; get delay parameter
8113  31 1B               acall  dly1                 ; wait for selected time
                          ;
                          ; move pointer and check for end of table
                          ;
8115  08                  inc    r0                   ; point to next pattern
8116  09                  inc    r1                   ; point to its delay time
8117  DA F6               djnz   r2,next              ; and repeat, if end of table
8119  80 EE               sjmp   cycle                ; start again
                          ;
                          ; time delay, number 0.5 secs passed via reg C
                          ;
811B  75 41 C8   dly1:    mov    count2,#t_500µs      ; initialize delay
811E  75 42 F8   dly2:    mov    count3,#t_2µs        ; counters 1 & 2
```

```
8121  D5 42 FD  dly3:      djnz     count3,dly3     ; count down 2 µs periods
8124  D5 41 F7             djnz     count2,dly2     ; count down 500 µs periods
8127  D5 40 F1             djnz     count1,dly1     ; count down 100 µms periods
812A  22                   ret                      ; go back to main program
                           ;
                           ORG      40h
                           ;
                           ; traffic light patterns and delay times
                           ;
0040  00        count1:    DB       0               ; counter for 100 ms periods
0041  00        count2:    DB       0               ; counter for 500 µs periods
0042  00        count3:    DB       0               ; counter for 2 µs periods
0043            ;
0043  41        step:      DB       green_ew + red_ns
0044  21                   DB       amber_ew + red_ns
0045  11                   DB       red_ew   + red_ns
0046  13                   DB       red_ew   + amber_ns + red_ns
0047  14                   DB       red_ew   + green_ns
0048  12                   DB       red_ew   + amber_ns
0049  11                   DB       red_ew   + red_ns
004A  31                   DB       amber_ew + red_ew   + red_ns
                           ;
004B  64 14 0A 14 delays:  DB       ten_sec,two_sec,one_sec,two_sec
004F  64 14 0A 14          DB       ten_sec,two_sec,one_sec,two_sec
                           ;
                           END      start
```

Note the use of a subroutine for the time delays in this program.

Subroutines

If an instruction sequence is to be used several times in a program, such as the time delays in Example 10, it is advantageous if the sequence is written so that it may be repeatedly called from any point within a program without having to rewrite it each time that it is used. An instruction sequence written this way is called a **subroutine** (or **procedure**), and its basic mechanism is shown in Figure 6.18.

Two instructions are provided in most instruction sets for implementing subroutines, and these are:

1 **CALL** which saves the current value of the program counter (the **return address**) and then reloads it with the address specified by the CALL operand, thus directing program execution to the **start of the subroutine**, and

2 **RET** which **recovers the return address** stored by the CALL instruction, thus causing program execution to continue from the instruction following the CALL instruction.

Both CALL and RET instructions may be **conditional** (i.e. depend upon the state of one of the status flags).

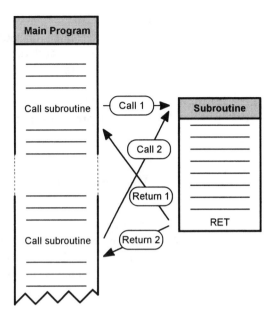

Figure 6.18

The return address is automatically saved in a section of RAM called the **stack** which operates as a **last-in-first-out** (LIFO) buffer, with a **stack pointer** (SP) register keeping track of where the last item was stored. This enables subroutines to be called from subroutines (**nested subroutines**), with each return address stored in the order that the subroutines were called, and recovered in the reverse order when returning from each subroutine. Registers may also be temporarily saved and restored by the use of PUSH and POP instructions.

A subroutine may require data to be passed to it; for example, when using a time delay subroutine it is useful if different time delays can be obtained by selecting different parameters before calling the subroutine. This is a technique known as **parameter passing**, and in its simplest form consists of storing the parameter in a register before calling the subroutine, then using that register during the subroutine. This is the technique used in Example 10.

Activity

1   A program is required to **display its program bytes**, as a sequence of binary codes, on the LEDs of an interface circuit of the type shown in Figure 6.8. Transfer of the next byte in the sequence is controlled by operating switch S0, and the program terminates once the last byte has been displayed.

2 Produce an **actions list**, **structure chart**, **pseudo-code** and program code required to implement the system described.

3 Using an appropriate microcomputer system, **test your program**, and check for correct operation.

Waveform generation

A microcomputer may be called upon to generate **analogue signals** of differing waveshapes and amplitudes rather than the purely digital outputs considered so far. This requires the use of a **digital to analogue** (DAC) interface, similar to that shown in Figure 6.19, to convert the two-level digital signals into multi-level analogue signals.

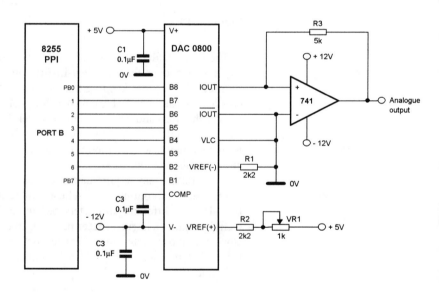

Figure 6.19

Depending upon the waveshape required, the digital inputs to the DAC may be generated by **mathematical processes** or by using **look-up tables**, both techniques having been dealt with in previous examples. Linear waveshapes such as ramp or triangular can be generated using add or subtract operations, but non-linear waveshapes such as sinewaves are probably best generated using look-up table techniques.

Example 11 – Waveform generation

> A program is required to generate a maximum amplitude ramp
> (sawtooth) waveshape of frequency 100 Hz, using the interface
> circuit shown in Figure 6.19.

The digital inputs to the DAC vary between 00000000_2 and 11111111_2,
therefore there are 256 discrete outputs, as shown in Figure 6.20.

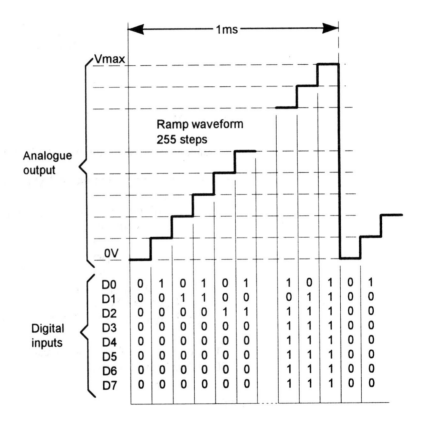

Figure 6.20

The time for one cycle of the 100 Hz ramp waveform is 1/100 or
10 ms, therefore the delay between output changes is 10/255 ms or
39.2 µs.

A structure chart for the ramp waveshape generator program is as shown
in Figure 6.21.

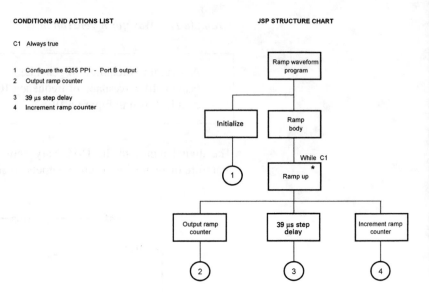

CONDITIONS AND ACTIONS LIST

C1 Always true

1 Configure the 8255 PPI - Port B output
2 Output ramp counter

3 39 μs step delay
4 Increment ramp counter

Figure 6.21

Pseudo-code for the ramp waveshape generator is as follows:

```
RAMP WAVEFORM PROGRAM  seq
     DO  1                        {configure 8255 PPI}
     RAMP BODY  iter  WHILE C1 {always}
          RAMP UP  seq
               DO  2             {output ramp counter to DAC}
               DO  3             {39 μs step delay}
               DO  4             {increment ramp counter}
          RAMP UP  end
     RAMP BODY  end
RAMP WAVEFORM PROGRAM  end
```

The following listings show typical assembly language programs for the
Z80, 8086 and 8031 MPUs.

Z80

```
; ******** Z80 Waveform generation *********
; PROGRAM 34
; generates 100 Hz ramp waveform via D to A
; converter attached to Port B.
; Assumes 2 MHz clock.
; Uses circuit shown in Figure 6.19
; ******************************************
;
0010 =      ppi       EQU     10h        ; 8255 PPI base address
0011 =      dac       EQU     ppi+1      ; D to A converter
0013 =      control   EQU     ppi+3      ; PPI control register
0090 =      iobyte    EQU     90h        ; I/O definition byte
```

```
                              ;
1800                          ORG      1800h          ; start of user RAM
                              ;
                              ; configure 8255 PPI
                              ;
1800  3E 90   start:          ld       a,iobyte       ; set up Port B as outputs
1802  D3 13                   out      (control),a    ;
                              ;
                              ; output ramp counter (initial value unimportant)
                              ; 255 steps for each cycle of the waveform
                              ;
1804  D3 11   ramp:           out      (dac),a        ; 11T    | output counter
                              ;
                              ; time delay for each output step -- uses time
                              ; wasting instructions to add 53T states to loop
                              ;
1806  DD E3                   ex       (sp),ix        ; 23T    | pad out loop to
1808  DD E3                   ex       (sp),ix        ; 23T    | give a step time
180A  06 00                   ld       b,0            ;  7T    | of 39 μs
                              ;
                              ; increment ramp counter and repeat
                              ;
190C  3C                      inc      a              ;  4T    | next step in
180D  C3 04 18                jp       ramp           ; 10T    | sequence
                                                      ; 78T x 0.5 μs = 39 μs
                                                      ; ---
                              ;
                              END      start
```

8086

```
                              ; ********* 8086 Waveform generation ********
                              ; PROGRAM 35
                              ; generates 100 Hz ramp waveform via D to A
                              ; converter attached to Port B.
                              ; Assumes 2 MHz clock.
                              ; Uses circuit shown in Figure 6.19
                              ; *****************************************
                              ;
                              .MODEL   SMALL
                              .STACK
                              ;
= 0010        ppi             EQU      10h            ; 8255 PPI base address
= 0011        dac             EQU      ppi+1          ; D to A converter
= 0013        control         EQU      ppi+3          ; PPI control register
= 0090        iobyte          EQU      90h            ; I/O definition byte
                              ;
                              .CODE
0400                          ORG      0400h          ; start of user RAM
                              ;
                              ; configure 8255 PPI
                              ;
0400  B0 90   start:          mov      al,iobyte      ; set up Port B as outputs
0400  E6 13                   out      control,al     ;
                              ;
                              ; output ramp counter (initial value unimportant)
                              ; 255 steps for each cycle of the waveform
                              ;
0404  E6 11   ramp:           out      dac,al         ; 10T    | output counter
                              ;
                              ; time delay for each output step -- uses time
                              ; wasting instructions to add 50T states to loop
                              ;
```

```
0406  90                      nop                         ;   3T   | pad out loop to
0407  B9 0000                 mov       cx,0              ;   4T   | give a step time
040A  B9 0003                 mov       cx,3              ;   4T   | of 39 µs
040D  E2 FE     dly:          loop      dly               ;  17T + 17T + 5T
                              ;
                              ; increment ramp counter and repeat
                              ;
040F  FE C0                   inc       al                ;   3T   | next step in
0411  EB F1                   jmp       ramp              ;  15T   | sequence
                              ;  ---
                              ;  78T x 0.5 µs = 39 µs
                              ;  ---
                              ;
                              END       start
```

                                      8031

```
                              ; ******** 8031 Waveform generation *********
                              ; PROGRAM 36
                              ; generates 100 Hz ramp waveform via D to A
                              ; converter attached to Port B.
                              ; Assumes 12 MHz clock (each machine cycle
                              ; takes 12 clock cycles, therefore T = 1 µs).
                              ; Uses circuit shown in Figure 6.19
                              ; *******************************************
                              ;
           ppi                EQU       0ff40h            ; 8255 PPI base address
           dac                EQU       ppi+1             ; D to A converter
           control            EQU       ppi+3             ; PPI control register
           iobyte             EQU       90h               ; I/O definition byte
                              ;
8100                          ORG       8100h             ; start of user RAM
                              ;
                              ; configure 8255 PPI
                              ;
8100  90 FF 43  start:        mov       dptr,#control ; set up Port B as outputs
8103  74 90                   mov       a,#iobyte         ;
8105  F0                      movx      @dptr,a           ;
                              ;
                              ; output ramp counter (initial value unimportant)
                              ; 255 steps for each cycle of the waveform
                              ;
8106  90 FF 41                mov       dptr,#dac         ;          | output ramp
8109  F0        ramp:         movx      @dptr,a           ;   2T     | counter
                              ;
                              ; time delay for each output step -- uses time
                              ; wasting instructions to add 34T states to loop
                              ;
810A  00                      nop                         ;   1T     | pad out loop
810B  78 24                   mov       r0,#36            ;   1T     | to give step
810D  D8 FE     dly:          djnz      r0,dly            ;   2T x 16 | time of 39 µs
                              ;
                              ; increment ramp counter and repeat
                              ;
810F  04                      inc       a                 ;   1T     | next step in
8110  80 F7                   sjmp      ramp              ;   2T     | sequence
                              ;  ---
                              ;  39T x 1 µs = 39 µs
                              ;  ---
                              ;
                              END       start
```

Activity

1 A program is required to generate a **triangular wave**, repetition rate **100 Hz**, at the output of an interface of the type shown in Figure 6.19.

2 Devise an **action list**, **structure chart**, **pseudo-code** and **program code** to enable the system to operate as described.

3 Using an appropriate microcomputer system, **test your program**, and check for correct operation.

Serial I/O

Many applications require **parallel data transfers** in which all data bits are transferred **simultaneously** along a parallel bus. Other applications require **serial data transfers** with data bits being transferred **one after another** along a single conductor. It therefore follows that conversions between parallel and serial data may be required, and may readily be accomplished by suitable software.

Microprocessor instruction sets include **shift** and **rotate** instructions which cause all bits in a register to move one place to the right or to the left according to the actual instruction used. Examples of these instructions are shown in Figure 6.22.

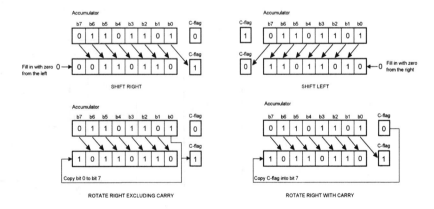

Figure 6.22

Parallel to serial and serial to parallel conversions may be achieved by a number of shifts with suitable delays between each shift, depending upon the serial data transfer rate.

Example 12 – Parallel to serial data transfer

An interface circuit similar to that shown in Figure 6.8 is connected to a microcomputer. A program is required to **read parallel data** from the switches connected to Port A, to **convert the data into serial form** with bit periods of 500 μs, and to transfer the serial data out via bit 0 of Port B (bit 0 first). The program must also generate a 100 μs pulse on bit 1 of Port B to indicate the start of each serial data transmission. The input switches are operated and a **double beam oscilloscope** is used to monitor bits 0 and 1 of Port B to observe the serial output data.

The operation of this system is shown in Figure 6.23.

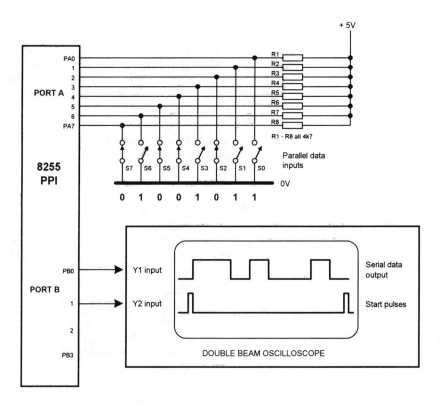

Figure 6.23

An action list and structure chart for the parallel to serial conversion program are shown in Figure 6.24.

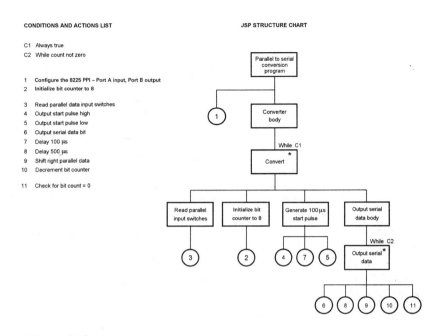

Figure 6.24

Pseudo-code for the parallel to serial conversion program is as follows:

```
PARALLEL TO SERIAL CONVERSION PROGRAM  seq
        DO   1                    {configure 8255 PPI}
        CONVERTER BODY  iter    WHILE  C1   {always}
             CONVERT seq
                     DO   3        {read parallel switches}
                     DO   2        {initialize bit count = 8}
                     DO   4        {generate start pulse rising edge}
                     DO   7        {delay 100 μs}
                     DO   5        {start pulse falling edge}
                     OUTPUT SERIAL DATA BODY  iter
                          OUTPUT SERIAL DATA  seq
                               DO  6    {output serial data bit}
                               DO  8    {delay 500 μs}
                               DO  9    {shift right parallel data}
                               DO  10   {decrement bit counter}
                               DO  11   {check for bit count = 0}
                          OUTPUT SERIAL DATA        end
                     OUTPUT SERIAL DATA BODY        end
             CONVERT  end
        CONVERTER BODY  end
PARALLEL TO SERIAL CONVERSION PROGRAM  end
```

The following listings show typical assembly language programs for the Z80, 8086 and 8031 MPUs.

Z80

```
                      ; ****** Z80 parallel to serial ******
                      ; PROGRAM 37
                      ; reads parallel input from switches on
                      ; Port A and converts to serial output
                      ; at Port B bit 0: assumes 2 MHz clock.
                      ; Uses circuit shown in Figure 6.23
                      ; *************************************
                      ;
0010 =      ppi       EQU    10h              ; 8255 PPI base address
0010 =      switch    EQU    ppi              ; parallel input switches
0011 =      serial    EQU    ppi+1            ; serial I/O port
0013 =      control   EQU    ppi+3            ; PPI control register
0090 =      iobyte    EQU    90h              ; I/O definition byte
                      ;
1800                  ORG    1800h            ; start of user RAM
                      ;
                      ; configure 8255 PPI
                      ;
1800 3E 90  start:    ld     a,iobyte         ; set up Port A as inputs
1802 D3 13            out    (control),a      ; and Port B as outputs
                      ;
                      ; read parallel inputs and initialize bit count
                      ;
1804 DB 10  conv:     in     a,(switch)       ; get parallel data
1806 4F               ld     c,a              ; and retain in C reg
1807 06 08            ld     b,8              ; count out 8 bits
                      ;
                      ; generate start bit pulse on bit 1 of Port B
                      ; delay time approximately 100 µs (200T)
                      ;
1809 3E 02            ld     a,2              ; set bit 1 of Port B
180B D3 11            out    (serial),a       ;
180D 1E 0D            ld     e,13             ;   7T        | 100 µs delay
180F 1D     dly100:   dec    e                ;   4T x 13 =   53T
1810 C2 0F 18         jp     nz,dly100        ;  10T x 13 =  130T
1813 AF               xor    a                ;   4T       | rest bit 1
1814 D3 11            out    (serial),a       ;  10T       | of Port B
                      ;
                      ; output serial data from bit 0 of Port B
                      ; loop time approximately 500 µs (1000T)
                      ;
1816 79     shift:    ld     a,c              ;   4T | time padding
1817 E6 01            and    l                ;   7T | instructions only
1819 79               ld     a,c              ;   4T | recover parallel bits
181A E6 01            and    l                ;   7T | mask off bits 1 to 7
181C D3 11            out    (serial),a       ;  11T | output serial data bit
181E 1E 41            ld     e,65             ;   7T | 500 µs delay
1820 1D     dly500:   dec    e                ;   4T x 67 = 268T | and wait
1821 C2 20 18         jp     nz,dly500        ;  10T x 67 = 670T |
1824 CB 39            srl    c                ;   8T | shift out next bit
1826 05               dec    b                ;   4T |
1827 C2 16 18         jp     nz,shift         ;  10T | and repeat for all 8 bits
                      ;
182A C3 04 18         jp     conv             ; get next parallel input
                      ;
                      END    start
```

8086

```
                         ; ****** 8086 parallel to serial ******
                         ; PROGRAM 38
                         ; reads parallel input from switches on
                         ; Port A and converts to serial output
                         ; at Port B bit 0: assumes 2 MHz clock.
                         ; Uses circuit shown in Figure 6.23
                         ; ************************************
                         ;
                         .MODEL    SMALL
                         .STACK
                         ;
= 0010      ppi          EQU       10h             ; 8255 PPI base address
=           switch       EQU       ppi             ; parallel input switches
= 0011      serial       EQU       ppi+1           ; serial I/O port
= 0013      control      EQU       ppi+3           ; PPI control register
= 0090      iobyte       EQU       90h             ; I/O definition byte
                         ;
                         .CODE
0400                     ORG       0400h           ; start of user RAM
                         ;
                         ; configure 8255 PPI
                         ;
0400  B0 90    start:    mov       al,iobyte       ; set up Port A as inputs
0402  E6 13              out       control,al      ; and Port B as outputs
                         ;
                         ; read parallel inputs and initialize bit count
                         ;
0404  E4 10    conv:     in        al,switch       ; get parallel data
0406  8A E0              mov       ah,al           ; and retain in C reg
0408  B3 08              mov       bl,8            ; count out 8 bits
                         ;
                         ; generate start bit pulse on bit 1 of Port B
                         ; loop time 100 µs (200T)
                         ;
040A  B0 02              mov       al,2            ; set bit 1 of Port B
040C  E6 11              out       serial,al       ;
040E  B9 000B            mov       cx,11           ; 4T | 100 µs delay
0411  E2 FE    dly100:   loop      dly100          ;(11 x 17T) + 5T = 192T
0413  B0 00              mov       al,0            ; 4T |
0415  E6 11              out       serial,al       ; reset bit 1 of Port B
                         ;
                         ; output serial data from bit 0 of Port B
                         ; loop time 500 µs (1000T)
                         ;
0417  8A C4    shift:    mov       al,ah           ; 2T | time padding only
0419  8A C4              mov       al,ah           ; 2T | recover parallel bits
041B  24 01              out       serial,al       ; 10T | output serial data bit
041F  B9 0037            mov       cx,55           ; 4T | 500 µs delay
0422  E2 FE    dly500:   loop      dly500          ;(56 x 17T) + 5T = 957T
0424  D0 EC              shr       ah,l            ; 2T | shift out next bit
0426  FE CB              dec       bl              : 3T | reduce bit count by 1
0428  75 ED              jne       shift           : 16T | and count 8 bits
                         ;
042A  EB D8              jmp       conv            ; get next parallel input
                         ;
                         END       start
```

8031

```
                         ; ****** 8031 parallel to serial ******
                         ; PROGRAM 39
                         ; reads parallel input from switches on
                         ; Port A and converts to serial output
                         ; at Port B bit 0:
                         ; assumes 12 MHz clock (1 µs cycle time).
                         ; Uses circuit shown in Figure 6.23
                         ; ***************************************
                         ;
                 ppi     EQU    0ff40h          ; 8255 PPI base address
                 switch  EQU    ppi             ; parallel input switches
                 serial  EQU    ppi+1           ; serial I/O port
                 control EQU    ppi+3           ; PPI control register
                 iobyte  EQU    90h             ; I/O definition byte
                         ;
8100                     ORG    8100h           ; start of user RAM
                         ;
                         ; configure 8255 PPI
                         ;
8100 90 FF 43  start:    movx   dptr,#control   ; set up Port A as inputs
8103 74 90               mov    a,#iobyte       ; and Port B as outputs
8105 F0                  movx   @dptr,a
                         ;
                         ; read parallel inputs and initialize bit count
                         ;
8106 90 FF 40  conv:     mov    dptr,#switch    ; pointer to switches
8109 E0                  movx   a,@dptr         ; get parallel data
810A F8                  mov    r0,a            ; and retain
810B 79 08               mov    r1,#8           ; count out 8 bits
                         ;
                         ; generate start bit pulse on bit 1 of Port B
                         ; delay time approximately 100 µs (200T)
                         ;
810D A3                  inc    dptr            ; point to serial output
810E 74 02               mov    a,#2            ; set bit 1 of Port B
8110 F0                  movx   @dptr,a         ;
8111 7A 30               mov    r2,#48          ; 1T  | 100 µs delay
8113 DA FE    dly100:    djnz   r2,dly100       ; 2T x 48 = 96T
8115 E4                  clr    a,              ; 1T  | reset bit 1
8116 F0                  movx   @dptr,a         : 2T  | of Port B
                         ;
                         ; output serial data from bit 0 of Port B
                         ; loop time approximately 500 µs (1000T)
                         ;
8117 E8                  mov    a,r0            ; 1T  | get parallel data
8118 54 01    shift:     anl    a,#1            ; 1T  | mask off bits 1 to 7
811A F0                  movx   @dptr,a         ; 2T  | output serial data bit
811B 00                  nop                    ; 1T  | time padding
811C 7B 05               mov    r3,#5           ; 1T  | 500 µs delay
811E 00       dly500:    nop                    ; 1T x 5       = 5T | time padding
811F 7A 2F               mov    r2,#47          ; 1T x 5       = 5T
8121 DA FE    d100:      djnz   r2,d100         ; 2T x 47 x 5 = 470T
8123 DB F9               djnz   r3,dly500       ; 2T x 5       = 5T
8125 E8                  mov    a,r0            ; 1T  | recover switch data
8126 03                  rr     a               ; 1T  | shift right one place
8127 F8                  mov    r0,a            ; 1T  | and save for next bit
8128 D9 EE               djnz   r1,shift        ; 2T  | repeat for all 8 bits
                         ;
812A 80 DA               sjmp   conv            ; get next parallel input
                         ;
                         END    start
```

Activity

1   Devise an **action list**, **structure chart, pseudo-code** and **program code** for a program to read input data as generated by a microcomputer using the program in Example 12.

2   Using an appropriate microcomputer system, **test your program**, and check for correct operation.

7 Interfacing microprocessor-based systems

Summary

A computer interacts with external devices as part of an entire system, and these devices are generally called **peripherals**; connecting peripherals to a microcomputer is a process known as **interfacing.** This element deals with some of the basic concepts of interfacing, followed by more detailed coverage of a number of interfacing topics, including parallel and serial I/O, code conversion between digital and analogue signals, and timing protocols such as polling and interrupts. Practical activities are suggested, and self-test exercises are included.

Interface concepts

The function of any computer is to accept data from the outside world, process that data and deliver results to the outside world. External devices or systems that provide a computer with its data are called **input peripherals**. External devices or systems that receive data from a computer are called **output peripherals**. Input and output peripherals are **rarely compatible** with a microcomputer bus system, therefore interfaces are required to overcome this incompatibility and ensure that peripherals can communicate effectively with the data bus. These interfaces are arranged as shown in Figure 7.1.

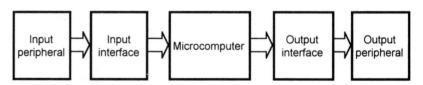

Figure 7.1

   An interface therefore allows two otherwise incompatible systems to be interconnected.

Synchronization of data transfers

For effective data transfers between peripheral devices and a microcomputer, data must be present in the **correct place** at the **exact time** expected. Since a microcomputer and its peripherals normally operate at

206

different speeds, some form of timing control is usually necessary, i.e. **synchronization**. Synchronization of data transfers may be achieved in a number of different ways, some of which are described in the following subsections.

Synchronous operation

In certain applications it may be possible to make a microcomputer and its peripheral operate in step with each other, i.e. **synchronously**, by using a **common clock** signal, as shown in Figure 7.2.

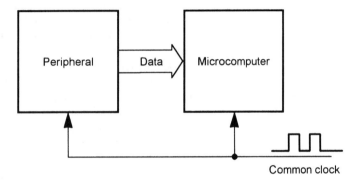

Figure 7.2

This is normally only advantageous when the two systems already operate at similar rates and it is more usual to use asynchronous operation.

Handshaking

Two control lines are used in addition to the data path, called **handshake lines**, organized as shown in Figure 7.3.

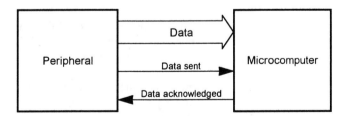

Figure 7.3

When new data is sent to a microcomputer by a peripheral device, one handshake line (DATA SENT or STROBE) is activated to inform the microcomputer that new data is available for processing. No further new data can be sent until a second handshake line (DATA ACKNOWLEDGED or READY) is activated to inform the peripheral that the previous data has been received and processed. When new data is sent to a peripheral by a microcomputer, the handshake lines operate in the reverse manner. Therefore, data transfers take place at the maximum rate possible, consistent with freedom from data loss or data duplication.

Buffers

Sometimes neither of the previous methods is suitable, particularly in cases where a microcomputer is unable to give immediate attention to the peripheral. An alternative method of transfer may be used in which a **buffer memory** is filled with data at a rate determined by the input system, and is emptied at a rate determined by the output system, e.g. filled by a microcomputer and emptied by a peripheral. The buffer fills and empties to absorb the speed differences between peripheral and microcomputer. Simple handshaking may also be needed to signify *buffer full* or *buffer empty* conditions, particularly if the buffer is rather small relative to the amount of data transferred. This method is often used for printer or keyboard interfaces and is illustrated in Figure 7.4.

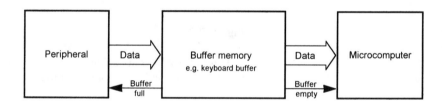

Figure 7.4

Code conversion

Microcomputers are **digital systems**, i.e. they process numbers, which for practical reasons must be binary. If input peripherals are capable of delivering digital signals, then they may be directly interfaced to a microcomputer input port. Similarly, if output peripherals controlled by a microcomputer are capable of accepting digital signals, then these may be directly interfaced to an output port. A purely digital system of this type is shown in Figure 7.5.

Since numerical data **changes abruptly**, in **discrete steps** from one value to the next, digital signals are **discontinuous** and intermediate values must therefore be represented by the nearest step in the sequence.

By way of contrast, physical changes of such quantities as temperature, pressure, velocity etc. have **infinite variations** and **change smoothly** from

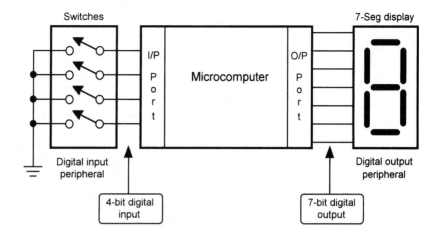

Figure 7.5

one value to the next. The electrical outputs from transducers that are used to measure physical quantities also change smoothly and are said to be analogous to the physical quantity, i.e. an **analogue system**.

When a computer system is required to process analogue signals, **code conversion** is required in order to change analogue signals into their digital equivalents or vice versa. This is achieved by using **analogue to digital** (ADC) or **digital to analogue** (DAC) converters, as shown in Figure 7.6.

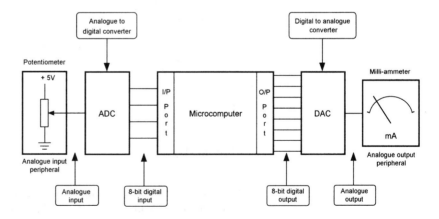

Figure 7.6

Serial/parallel conversion

Data are processed within a microcomputer as groups of 8, 16 or 32 bits depending upon the type of MPU used and the number of conductors forming the data bus. Since at a given instant these data bits are all present on the data bus, it is called **parallel** data. Many computer peripherals generate parallel codes, or expect to receive them,

e.g. a keyboard may generate 7-bit parallel (ASCII) code, and a 7-segment display is driven by a 7-bit parallel code.

Sometimes it is not convenient, or not possible, to transfer all bits of a parallel code simultaneously. This may be due to the fact that there are insufficient data channels available, or simply that the distance over which the transfer is to take place is too great for reliable parallel transfers. If a data path is narrowed by reducing the number of data lines it contains, then it is no longer possible for the same number of bits to be transferred along it in parallel, and some bits must follow on later. Taking this to its logical conclusion, a data path could be narrowed to just one bit and 8-bit parallel data could only be transferred along it one bit at a time, as a procession of 8 bits in time sequence. This **bit stream** is commonly known as **serial** data and if used for data transfers, requires parallel data to be converted into serial form prior to transmission, and converted back into parallel form when received. An arrangement of this type is shown in Figure 7.7.

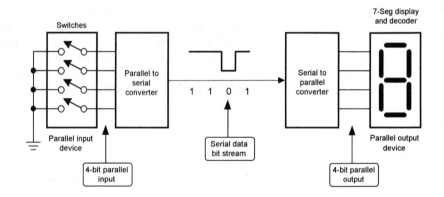

Figure 7.7

Multiplexing

The number of I/O lines available on a microcomputer is limited, often only two 8-bit ports with 16 I/O lines. Sometimes this is insufficient and it becomes necessary to operate a system such that two or more peripherals share the same I/O lines. This technique is called **multiplexing** and is really no different to several devices within a microcomputer sharing a common data bus.

Obviously it is not possible for two peripheral devices to share an I/O line and expect it to simultaneously transfer different data, therefore each peripheral must be allocated a particular time slot and this is a technique known as **time division multiplexing** (TDM).

The principal advantage of multiplexing is that for a given application it can result in a reduction in the number of data I/O lines needed (but at the expense of a lower data transfer rate). The use of multiplexing to reduce the number of I/O lines (and other components) needed when driving 7-segment displays is shown in Figure 7.8.

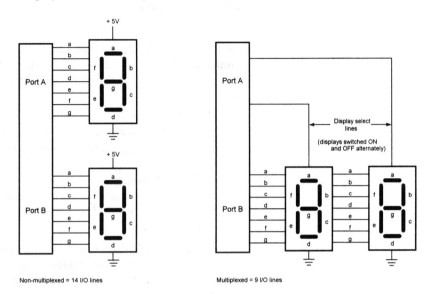

Non-multiplexed = 14 I/O lines   Multiplexed = 9 I/O lines

Figure 7.8

Electrical conversion

This process involves changing the electrical nature of input or output signals as they are transferred between a microcomputer and its peripherals in order to maintain compatibility.

Signal conditioning

The signals obtained from input transducers are rarely suitable for direct processing by a microcomputer system. For example, a transducer may not actually generate an electro-motive force, but may change in resistance or capacitance etc., in response to changes in physical conditions being measured.

Those that do generate emfs may do so at too low or too high a level for direct use. Signal conditioning may therefore consist of amplification, attenuation, '**shaping**' or '**cleaning**' of signals before or after processing. Hardware alone may be used to carry out this process, but often a combination of both hardware and software is used. Conditioning of signals prior to processing is known as **pre-conditioning**, while conditioning after processing is called **post-conditioning**. Often hardware is used for pre-conditioning and software is used for post-conditioning.

Electrical isolation

Microprocessor-based systems are often interfaced to mains operated equipment, or other high voltage systems. Direct connection between a microcomputer and such equipment should be avoided for the following reasons:

1 **Operator safety**
   Due to fault conditions (or other reasons) an operator could come into contact with mains potentials with possible lethal results.

2 **Equipment damage**
   Microcomputer components are designed to operate at low d.c. potentials and would therefore be damaged if coming into contact with high potentials.

Therefore **electrical isolation** is essential (typically 2–3 kV), and is often implemented using an **opto-isolator**. This is a device that contains a light source and a light-activated electronic switch in one package, organized as shown in Figure 7.9.

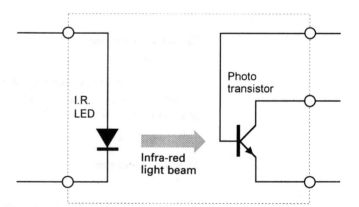

Figure 7.9

Current passing through the LED causes it to emit **infra-red light** which acts upon the base of a photo-transistor, causing it to conduct. When current flow through the LED is interrupted, the infra-red light is cut off, causing the photo-transistor to cease conducting. The photo-transistor may be used to switch current in the output circuit, but since there is **no direct electrical connection** between input and output of an opto-isolator, complete isolation is achieved.

Test your knowledge 7.1

1   An interface allows two otherwise:
    A   compatible systems to be interconnected
    B   incompatible systems to be isolated
    C   compatible systems to be isolated
    D   incompatible systems to be interconnected

2   A moving coil meter is:
    A   an analogue device since it changes in steps
    B   a digital device since it changes in steps
    C   an analogue device since it changes smoothly
    D   a digital device since it changes smoothly

3   A potentiometer must be connected to a micro-
    computer using:
    A   a DAC because it is an analogue device
    B   a DAC because it is a digital device
    C   an ADC because it is a digital device
    D   an ADC because it is an analogue device

4   A 7-segment LED is:
    A   an analogue device with an infinite number of
        different displays
    B   a digital device with a fixed number of different
        displays
    C   an analogue device with a fixed number of
        different displays
    D   a digital device with an infinite number of different
        displays

5   Parallel data transfers may be used between
    microcomputers and their peripherals using:
    A   a single data line up to 4 metres in length
    B   a single data line up to 30 metres in length
    C   multiple data lines up to 30 metres in length
    D   multiple data lines up to 4 metres in length

6   Serial data transfers are:
    A   faster than parallel but need only one data
        conductor
    B   faster than parallel but use the same number of
        data conductors
    C   slower than parallel but need only one data
        conductor
    D   slower than parallel but use the same number of
        data conductors

7 An opto-isolator is used between a microcomputer and peripheral to provide:
   A signal coupling and electrical isolation
   B signal isolation and electrical coupling
   C signal and electrical coupling
   D signal and electrical isolation

8 Multiplexing of two 7-segment display LEDs allows them to operate:
   A simultaneously but uses more port I/O lines
   B in time sequence but uses more port I/O lines
   C simultaneously but uses fewer port I/O lines
   D in time sequence but uses fewer port I/O lines

Parallel data I/O

Parallel inputs and outputs involve **simultaneous transfers** of two or more data bits via an I/O port. In practice a single I/O port normally has at least eight bits, and a microcomputer may have two or more 8-bit I/O ports. Each of these ports may be used for the transfer of a group of related bits, e.g. an 8-bit output to a DAC, or the port may transfer eight individual unrelated bits for control purposes (see Figures 7.10(a) and (b)).

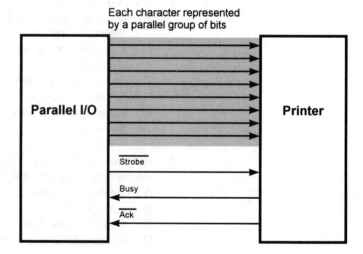

Figure 7.10(a)

I/O timing

All I/O transfers take place between peripheral devices and the MPU data bus with the main interfacing problem being one of **correct timing**, i.e. identifying when the bus holds valid data for an output port, or is ready to accept data from an input port.

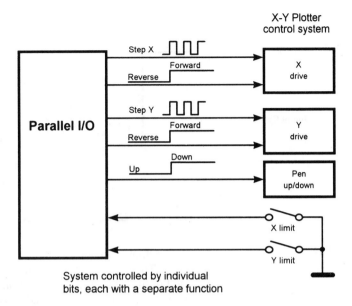

Figure 7.10(b)

An input peripheral needs an interface circuit that allows it to be electrically connected to the data bus only during execution of the relevant peripheral **read** instruction. An output peripheral needs an interface circuit that allows data to be captured from the data bus during execution of the relevant peripheral **write** instruction. Peripheral read and write instructions generate unique bit patterns on the address and control buses, and these are identified by a decoder circuit that generates control signals to establish communication between peripheral and data bus. The timing of these transfers may be studied by reference to appropriate timing diagrams.

Some MPUs have separate *(isolated)* I/O space that requires the use of special input and output instructions, e.g. IN or OUT, while other MPUs have **memory mapped I/O** which uses memory transfer instructions such as MOV or LD. All MPUs can use memory mapped I/O, but isolated I/O is reserved for certain MPUs such as Z80 or 8086, which have IN and OUT instructions and relevant control signals to switch between memory and I/O.

Data output latch

Data latches are basically simple digital storage devices, each having an **input terminal (D), an output terminal (Q)** and a **clock input (CK)**. With the clock input inactive, D is isolated from Q and changes at its input do not affect Q. Once the clock input changes to an active state, then Q follows any changes on D, and this continues until the clock returns to its inactive state. The Q terminal therefore stores the state on D at the instant the clock becomes inactive and this may be used as a means of capturing data from the MPU data bus, as shown in Figure 7.11.

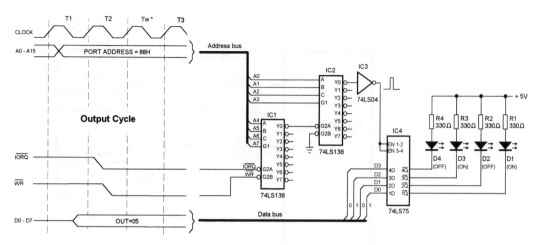

Figure 7.11

The output peripheral in this case is a group of LEDs (or opto-isolator inputs). The MPU places data for the LED onto the data bus for a short period of time during execution of an OUT instruction. The MPU also puts the port address (88h) onto the address bus, and takes $\overline{WR}$ and $\overline{IORQ}$ to their active states. I/O address decoders IC1 and IC2 monitor the address and control lines for these conditions, and when detected cause Y0 of IC2 to change state to logical 0 until the conditions are changed. A negative going pulse is therefore generated, and after inversion by IC3, is applied to the clock input of IC4. Data bus bits D0–D3 are applied to the data inputs of IC4, which in turn causes data outputs $\overline{Q0}$–$\overline{Q3}$ to assume the same data in inverted form during the positive clock period. At the instant IC4 clock input returns to logical 0, Q0–Q3 output data is latched and remains stable until arrival of the next clock pulse. The use of inverted outputs $\overline{Q0}$–$\overline{Q3}$ ensures that a logical 1 applied to the latch causes the corresponding LED to switch on.

Similar circuits to this are already built into the 8031 MPU and may be controlled by instructions such as the following:

clr p1.0	or	mov p1.0,#0	(clear bit 0 of port 1)
setb p1.2	or	mov p1.2,#1	(set bit 2 of port 1)

The advantage of using **mov** instructions rather that **clr** or **setb** is that several bits may be controlled simultaneously.

Activity

1 Study the technical data for any available micro-computer system.

2 Devise a circuit, similar to that shown in Figure 7.11, for an 8-bit parallel output port using a **74LS373 octal latch**, locating the output port at any free memory or I/O address.

3 Construct your circuit on breadboard, connect to the microcomputer buses, and test for correct operation.

4 Write suitable software to enable the output port to control a parallel peripheral such as a 7-segment display, DAC or stepper motor.

5 Write a report on your parallel output port, including circuit diagram, description of operation and software for controlling a peripheral.

Data input latch

An input peripheral cannot be connected permanently to the data bus since this could cause **bus conflicts** that would interfere with instruction fetch–execute processes. A peripheral must therefore be connected to the data bus through a **tri-state buffer** which is only enabled when the data bus is ready to receive data, i.e. during execution of an IN instruction. A suitable circuit is shown in Figure 7.12.

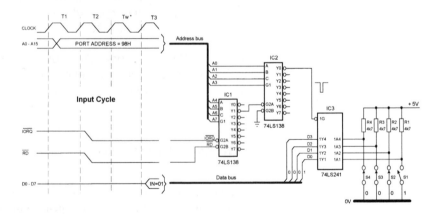

Figure 7.12

In this case the input peripheral is a bank of four switches. During execution of an IN instruction, the MPU places the port address (98h) onto the address bus and takes RD and IORQ to their active states. I/O address decoders IC1 and IC2 monitor the address and control lines for these conditions, and when detected cause Y0 of IC2 to change state to logical 0 until the conditions are changed. The negative going pulse generated as a result is used to enable tri-state buffers in IC3 which, in turn, connect switches S1–S4 to the data bus. The MPU clocks in data from the data bus during T3, thus reading the switches. After execution of the IN instruction, IC3 is again disabled and the switches are isolated from the data bus.

Similar circuits to this are already built into the 8031 MPU and may be controlled by the following instructions:

```
mov a,p1          (read port 1 into accumulator)
mov r0,p2         (read port 2 into register r0)
```

Activity

1   Study the technical data for any available micro-computer system.

2   Devise a circuit, similar to that shown in Figure 7.12, for an 8-bit parallel input port using a **74LS373 octal latch** (with tri-state outputs), locating the output port at any free memory or I/O address.

3   Construct your circuit on breadboard, connect to the microcomputer buses, and test for correct operation.

4   Write suitable software to enable the input port to read a parallel peripheral such as an 8-bit DIL switch, matrix keyboard or ADC.

5   Write a report on your parallel input port, including circuit diagram, description of operation and software for reading a peripheral.

Programmable I/O devices

Although interfacing may be achieved by the use of **latches** and **buffers**, the task may be simplified by using **programmable I/O devices**, a number of which are available, manufactured specifically for this purpose. The programmable nature of such devices allows them to be configured in a number of different ways, and permits ports to be specified as inputs or outputs without the need to make hardware changes.

8255 programmable peripheral interface

One device commonly used in many systems is the **Intel 8255 PPI**, no doubt due to the simplicity with which it can be interfaced to most MPU buses. The 8255 is a programmable peripheral interface (PPI) device that may be interfaced directly to most MPUs, providing a total of three 8-bit parallel I/O ports (Port A, Port B and Port C). The block and pin out diagrams of an 8255 PPI are shown in Figure 7.13.

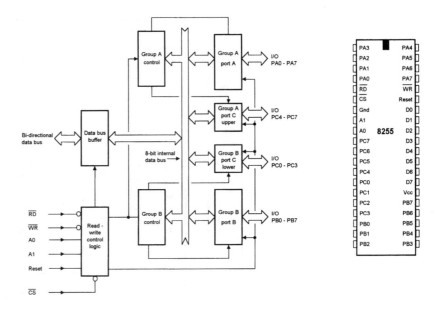

Figure 7.13

The 24 I/O pins may be programmed in two groups of 12 to form various combinations of 4- and 8-bit parallel ports depending upon the operating mode selected.

Operating modes

The 8255 may be operated in any of the following three modes:

1 Mode 0: basic I/O mode
2 Mode 1: strobed I/O mode
3 Mode 2: bidirectional I/O mode

Mode 0: This mode provides **simple I/O operations** without handshake facilities, data being simply written to or read from each of the three I/O ports as required. Ports A and B operate as two 8-bit ports, while Port C is operated as two 4-bit ports, and any port may be configured as either an input port or an output port.

Mode 1: This mode provides **I/O operations with handshake** facilities. Ports A and B may be defined as either input or output ports, and handshaking is provided by Port C. Six bits of Port C are used for handshaking and interrupt control, three bits for Port A and three bits for Port B.

Mode 2: This mode provides **bidirectional I/O** operations on Port A only. Handshaking is provided by five bits of Port C.

These modes of operation are summarized in Figure 7.14.

	MODE 0		MODE 1		MODE 2
	IN	OUT	IN	OUT	Group A only
PA0	IN	OUT	IN	OUT	←→
PA1	IN	OUT	IN	OUT	←→
PA2	IN	OUT	IN	OUT	←→
PA3	IN	OUT	IN	OUT	←→
PA4	IN	OUT	IN	OUT	←→
PA5	IN	OUT	IN	OUT	←→
PA6	IN	OUT	IN	OUT	←→
PA7	IN	OUT	IN	OUT	←→
PB0	IN	OUT	IN	OUT	—
PB1	IN	OUT	IN	OUT	—
PB2	IN	OUT	IN	OUT	—
PB3	IN	OUT	IN	OUT	—
PB4	IN	OUT	IN	OUT	—
PB5	IN	OUT	IN	OUT	—
PB6	IN	OUT	IN	OUT	—
PB7	IN	OUT	IN	OUT	—
PC0	IN	OUT	INTR B	INTR B	I/O
PC1	IN	OUT	$\overline{\text{IBF B}}$	$\overline{\text{OBF B}}$	I/O
PC2	IN	OUT	$\overline{\text{STB B}}$	$\overline{\text{ACK B}}$	I/O
PC3	IN	OUT	$\overline{\text{INTR A}}$	INTR A	INTR A
PC4	IN	OUT	$\overline{\text{STB A}}$	I/O	$\overline{\text{STB A}}$
PC5	IN	OUT	IBF A	I/O	IBF A
PC6	IN	OUT	I/O	$\overline{\text{ACK A}}$	$\overline{\text{ACK A}}$
PC7	IN	OUT	I/O	$\overline{\text{OBF A}}$	$\overline{\text{OBF A}}$

(PB rows for Mode 2: Mode 0 or Mode 1 only)

INTR = Interrupt

IBF = Input Buffer Full

$\overline{\text{OBF}}$ = Output Buffer Full

$\overline{\text{STB}}$ = Strobe

$\overline{\text{ACK}}$ = Acknowledge

Figure 7.14

An 8255 PPI may be interfaced to different MPUs as shown in Figures 7.15(a) and (b), and similar circuits may be used for most other MPUs.

Configuring an 8255 PPI

In common with all programmable devices, an 8255 PPI must first be instructed how it is to operate before data transfers can be effected between an MPU and its I/O ports. This is a process called **configuring**. The 8255 PPI has two port select signals (A0 and A1) which, in conjunction with the $\overline{\text{RD}}$ and $\overline{\text{WR}}$ inputs, control selection of one of the three I/O ports or the

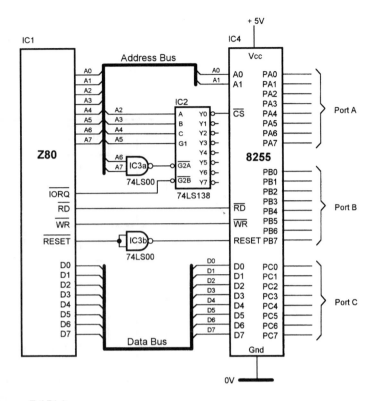

Figure 7.15(a)

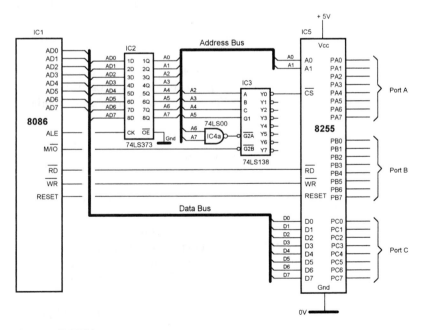

Figure 7.15(b)

control word register. Selection of an I/O port or the control word register is achieved by using an appropriate address in conjunction with an IN or OUT instruction (or MOV in memory mapped I/O systems), and the least significant two bits of the address identify the port or control register, as shown in Figure 7.16.

A1	A0	$\overline{RD}$	$\overline{WR}$	$\overline{CS}$	Input operation (READ)
0	0	0	1	0	Port A → Data bus
0	1	0	1	0	Port B → Data bus
1	0	0	1	0	Port C → Data bus
					Output operation (WRITE)
0	0	1	0	0	Data bus → Port A
0	1	1	0	0	Data bus → Port B
1	0	1	0	0	Data bus → Port C
1	1	1	0	0	Data bus → Control
					Disable function
X	X	X	X	1	Data bus → 3-state
1	1	0	1	0	Illegal condition

Figure 7.16

The contents of the control word register determine the mode selection and port configuration, as shown in Figure 7.17. Therefore, configuring an 8255 PPI involves storing an appropriate control word in this register.

Configuring example

An 8255 PPI is to be operated in Mode 0 as follows:

Port A input mode
Port B output mode
Port C (low) output mode
Port C (high) input mode

The control word required to configure the 8255 in this manner is 98H, as shown in Figure 7.18.

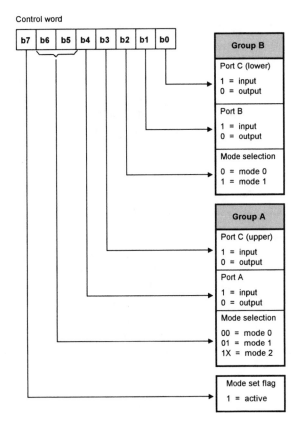

Figure 7.17

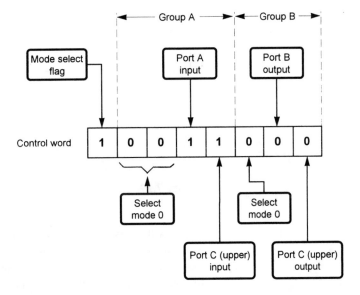

Figure 7.18

An instruction sequence similar to the following may be used to perform the actual configuring:

(a)		(b)		(c)	
Z80		8086		8031	
ld	a,98h	mov	a,98h	mov	dptr,#ff43h
out	(13h),a	out	13,a	mov	a,#98h
				movx	@dptr,a

Single bit control

For normal mode selection and configuring operations on an 8255 PPI, bit 7 of the control word register is set to logical 1. This bit behaves as a **mode set flag** and ensures that the data sent to the control address is stored in the **mode definition register** of the PPI. If a control word with bit 7 reset to 0 is used, a different function is provided at the control address. This function enables any of the eight bits of Port C to be set or reset using a single OUT instruction, as shown in Figure 7.19.

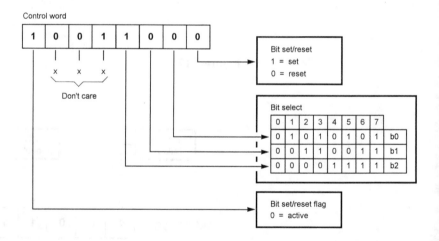

Figure 7.19

When the 8255 PPI is used in **mode 1** or **mode 2**, control signals are provided that may be used as **interrupt requests** for the MPU. The interrupt request signals generated by Port C, bits 0 and 3, may be enabled or disabled by setting or resetting the appropriate **INTE flip-flop** using the bit set/reset facility of Port C.

Status word

When operating in Mode 0, Port C is used for **peripheral data transfers**. When using modes 1 or 2, however, Port C **generates** or **accepts handshaking signals** for data transfers via Ports A and B. Reading Port C enables the status of each peripheral to be tested. The status word format for Modes 1 and 2 is shown in Figure 7.20.

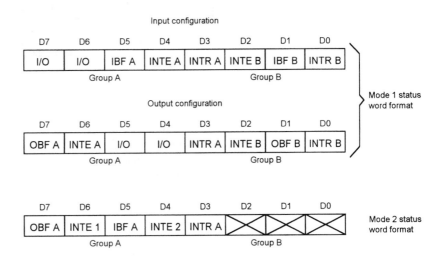

Figure 7.20

Activity

1   Study the technical data for any available micro-computer system.

2   Identify any programmable parallel I/O port device used.

3   Obtain a manufacturer's data sheet for the device identified.

4   Write a report on your parallel port, including a circuit diagram, details of the interfacing device, description of the interfacing protocols, and an evaluation of the software used.

Serial data I/O

The term **serial** when applied to data transfers usually means **bit-serial** in which the individual bits of data bytes are transferred in time sequence, one after another, as shown in Figure 7.21.

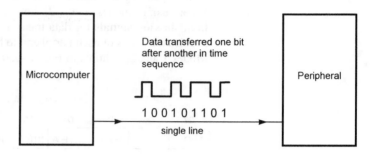

Figure 7.21

Serial data transfers are commonly used when peripheral devices are located at some distance from the microcomputer to which they are attached, since parallel transfers are limited to maximum distances of approximately **4 metres**. A typical arrangement for serial data communications is shown in Figure 7.22.

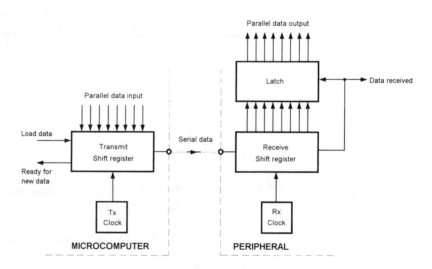

Figure 7.22

It can be seen from this diagram that the hardware required is relatively simple and cost-effective, requiring only one wire or data channel (plus ground) for each data path. This has the advantage that connections between a peripheral (e.g. keyboard or printer) and a microcomputer can be made without resorting to bulky multi-core or ribbon cables, and also avoids problems associated with interaction between adjacent signals when using long parallel conductors.

Asynchronous transfers

Data transfers usually take place **asynchronously** so that it is not necessary for the MPU and peripheral to share a common clocking signal (although both clocks must operate at very similar rates if errors are to be avoided). In effect, each data word transferred carries its own synchronizing signal in the form of **start** and **stop** bits which are appended as shown in Figure 7.23.

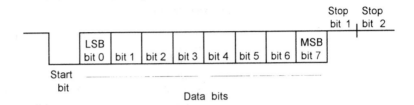

Figure 7.23

When in the receiving mode, a peripheral waits for the falling edge of each start bit before clocking in the following serial data, therefore discrepancies in the clocking frequency are acceptable provided that clocking still occurs within the correct bit period. A progressive shift in clocking point is corrected when the next start bit appears, thus preventing accumulative effects from corrupting received data (see Figure 7.24).

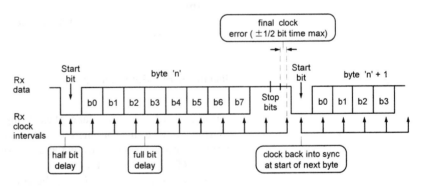

Figure 7.24

Baud rate

Since not all bits transferred in a serial system are actual data bits, expressing a data transfer rate in terms of **bits per second** may cause confusion. For example, if eight-bit data words are transferred at the rate of one word per second, the data transfer rate might be considered as eight bits per second. However, since each word has a start and stop bit added, in order to transfer one word per second, the transfer rate must be ten bits per second. This confusion is avoided by stating data transfer rates in terms of their **baud rate**.

The baud rate is defined as follows:

$$\text{Baud rate} = \frac{1}{\text{bit time}} \quad \text{(time of shortest element)}$$

Thus at **2400 baud**, the bit time is **1/2400** or **417 μs**. Consider the following two transfer formats:

	(a)	(b)
Data bits	7	6
Start bit	1	1
Stop bits	2	1
Total bits	10	8

Assuming **2400 baud**, the data transfer rates are:

(a)

$$\frac{10^6}{10 \times 417} = \textbf{240 characters/second}$$

(b)

$$\frac{10^6}{8 \times 417} = \textbf{300 characters/second}$$

Serial/parallel data conversion

Serial data transfers may take place using a **single I/O line** of a parallel interface device in conjunction with suitable software to perform the parallel/serial conversion, or by the use of **dedicated serial I/O controllers**, e.g. UART, USART, ACIA or SIO.

The first of these methods may be considered as implementing a UART in software and requires virtually no additional hardware. The second method does require additional hardware, but after configuring the device, little extra in the way of software.

Software UART

This arrangement uses software to simulate the functions of a **UART (Universal Asynchronous Receiver/Transmitter)**, and serial data is transferred via a single port line of a parallel I/O device with an MPU register acting as a shift register. It is therefore readily adaptable for different baud rates, and in many cases, automatically adjusts itself to the correct rate after measuring the baud rate of incoming data.

The hardware required for this system and a sample structure chart for parallel to serial conversion are shown in Figures 7.25(a) and (b).

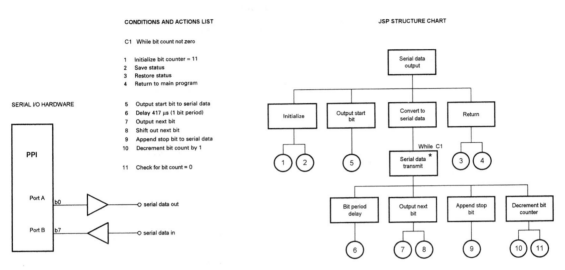

Figure 7.25(a)   **Figure 7.25(b)**

The software required depends upon the actual format used. The use of start and stop bits has already been discussed, and these must be added to data by the software prior to transmission. The format shown in Figure 7.25(b) is for **8-bit serial data**, **one start** bit and **two stop** bits. These extra bits must also be detected and removed by the software used for reception.

Pseudo-code for parallel to serial software is as follows:

```
SERIAL DATA OUTPUT   seq
        DO      1                    {initialize bit count to 11}
        DO      2                    {save status of registers}
        DO      5                    {output a start bit}
        CONVERT TO SERIAL DATA  iter  WHILE C1  {count not = 0}
                SERIAL DATA TRANSMIT   seq
                        DO      6            {delay 417 μs bit period}
                        DO      7            {output next bit}
                        DO      8            {shift out next bit}
                        DO      9            {append stop bit}
                        DO      10           {decrement bit counter by 1}
                        DO      11           {check for bit counter = 0}
                SERIAL DATA TRANSMIT   end
        CONVERT TO SERIAL DATA   end
        DO      3                    {restore status of registers}
        DO      4                    {return to main program}
SERIAL DATA OUTPUT   end
```

A sample assembly language listing for the 8086 MPU follows. This program may be simply adapted for other types of MPU and also for different baud rates.

```
; ********** 8086 software UART *********
; PROGRAM 40
; Subroutine to convert parallel data in
; register AL to serial and output via
; Port A bit 0 at 2400 baud.
; Assumes 2 MHz clock and PPI configured.
; Uses circuit shown in Figure 7.25(a)
; ****************************************
;
          .MODEL      SMALL
          .STACK
;
= 0010    sio    EQU    10h        ; serial O/P port (bit 0)
= 000B    bits   EQU    11         ; 8 data, 1 start and 2 stop
= 002C    t_bit  EQU    2ch        ; bit time for 2400 baud
;
          .CODE
0400      ORG    0400h             ; start of user RAM
;
; initialize bit count and send start bit
;
0400 53        start:  push   bx          ; save status of bx
0401 51                push   cx          ; and cx
0402 B3 0B             mov    bl,bits      ; count out 8 bits
0404 F8                clc                 ; start bit = 0
0405 D0 D0             rcl    al,1         ; move start bit to al
0407 E6 10     send:   out    sio,al       ; and transmit
;
; send 2400 baud data plus 2 stop bits
;
0409 B9 002C           mov    cx,t_bit     ; one bit time delay
040C E2 FE     dly1:   loop   dly1         ; of 417 µs
040E D0 D8             rcr    al,1         ; shift first bit to b0
0410 F9                stc                 ; set carry for stop bits
0411 FE CB             dec    bl           ; reduce bit counter by 1
0413 75 F2             jnz    send         ; and check for last bit
;
0415 59                pop    cx           ; restore status of bx
0416 5B                pop    bx           ; and cx
0417 CB                retf                ; return
;
          END    start
```

Hardware UART

Dedicated serial I/O controllers are available in most MPU families, each performing similar functions. Some UARTs are available that are purely hardware configured, but most now require some sort of software configuration. Certain MPUs, e.g. the 8085, have simple serial I/O facilities and most single chip microcontrollers, e.g. the 8031, have more comprehensive built-in serial I/O facilities.

8251 USART

One example of a serial I/O controller is the **8251 USART (Universal Synchronous/Asynchronous Receiver/Transmitter)**, and a block diagram of this device is shown in Figure 7.26.

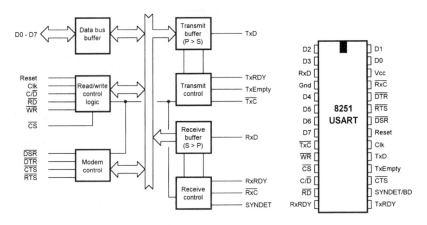

Figure 7.26

The 8251 contains separate **transmitter** (parallel to serial) and **receiver** (serial to parallel) sections which may be operated independently. Each section is capable of operating in **synchronous** or **asynchronous** modes, although the asynchronous mode is more likely to be used.

The 8251 is divided into two logical sections, **data** and **mode/control/status**, selected by the logic level applied to the C/$\overline{\text{D}}$ input (0 = data, 1 = mode/control/status). The C/$\overline{\text{D}}$ input is connected to an address line (usually A0 or A1) so that the two sections are assigned different addresses. The mode/control/status section is used to set up the serial data format, issue various commands and to determine the status of the device. The data section is used for both **transmit** (write) and **receive** (read) operations.

A data byte may be written to the transmitter section, if enabled, and is then immediately converted into serial form and sent out via the **TxD output**, using the specified format. The MPU may determine when the transmitter is available to send a character by monitoring **bit 0** of the **status register**. Provided the receiver is enabled, serial data applied to the **RxD input** is converted into parallel form and stored in the receiver buffer. The MPU is able to determine when a character has been completely received by reading **bit 1** of the **status register**.

The 8251 USART is programmed by performing the following operations after system reset:

1 **Select the mode register**
2 **Set up the serial word format**

The command register is automatically selected after setting up the serial word format. The function of each of the mode/command/status register bits is shown in Figure 7.27.

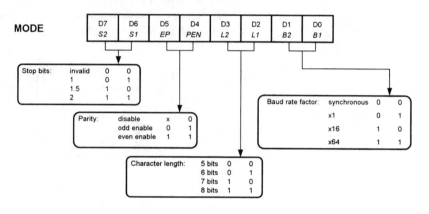

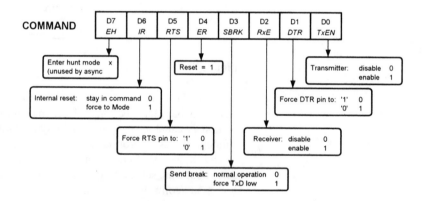

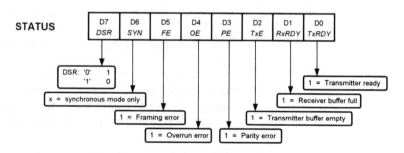

Figure 7.27

The following program code shows how an 8251 USART may be configured.

```
                           ; ***** 8086 USART configuring *****
                           ; PROGRAM 41
                           ; Routine to configure 8251 USART
                           ; needs delay after each configuring
                           ; operation if handshaking not used
                           ; ********************************
                           ;
                           .MODEL    SMALL
                           .STACK
                           ;
= 0020         usart       EQU       20h           ; usart Tx/Rx data buffer
= 0021         usartc      EQU       usart+1       ; mode/command/status reg.
                           ;
                           .CODE
0400                       ORG       0400h         ; start of user RAM
                           ;
                           ; USART internal reset, select 'COMMAND' register
                           ;
0400 B3 04     start:      mov       bl,4          ; internal reset,
0402 32 C0                 xor       al,al         ; repeat several times to
0404 E6 21     commd:      out       usartc,al     ; ensure that USART command
0406 FE VB                 dec       bl            ; register is selected
0408 75 FA                 jnz       commd         ;
                           ;
                           ; select USART 'MODE' register and set up serial
                           ; data format, i.e. number of data, stop and parity
                           ; bits and baud rate factor (clock multiplication)
                           ;
040A B0 40                 mov       al,40h        ; force to MODE
040C E6 21                 out       usartc,al     ; (set bit 6 in COMMAND reg.)
                           ;
                                                   ; MODE register = 4eh
                                                   ; b7 = 0   1 stop bit
040E B0 4E                 mov       al,4eh        ; b6 = 1
0410 E6 21                 out       usartc,al     ; b5 = 0   no parity
                                                   ; b4 = 0
                                                   ; b3 = 1   8 data bits
                                                   ; b2 = 1
                                                   ; b1 = 1   Tx/Rx clock 16x
                                                   ; b0 = 0   baud rate
                           ;
                           ; auto reverts to COMMAND register after setting MODE
                           ; set up operation of USART by initializing COMMAND reg.
                           ;
                                                   ; COMMAND register = 37h
0412 B0 37                 mov       al,37h        ; b7 = 0   ignore
0414 E6 21                 out       usartc,al     ; b6 = 0   stay in COMMAND
                                                   ; b5 = 1   force RTS pin low
                                                   ; b4 = 1   force USART to 'idle'
                                                   ; b3 = 0   send break normal op.
                                                   ; b2 = 1   enable receiver
                                                   ; b1 = 1   force DTR pin low
                                                   ; b0 = 1   enable transmitter
                                                   ;
                           ; read STATUS register at address 'usartc' to determine
                           ; when to transfer data via the Tx/Rx buffer at address
                           ; 'usart'
                           ;
                           END       start
```

RS-232 serial communications

The Electronic Industries Association (EIA) Recommended Standard (RS) **RS-232** contains a specification for the interconnection between two systems to allow them to communicate in serial fashion. This standard defines the permissible **voltage levels**, control lines and **physical connectors**, but does not concern itself with such items as number of bits, baud rate and other factors related to the speed of the serial data.

Most peripheral devices such as printers, VDUs or keyboards are available with RS-232 compatible I/O and this simplifies the connection of such peripherals to a microcomputer. Typically peripheral devices may be located at distances of up to **30 metres** when using this standard.

EIA RS-232 standards

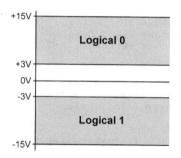

Figure 7.28

In order to minimize the effects of **electrical noise** upon serial data signals, the RS-232 standard specifies voltage levels of between **+3 V** and **+25 V** for **logical 0**, and between **−3 V** and **−25 V** for **logical 1**. The later RS-232C standard is normally used nowadays in which the voltage levels are reduced to + or − 15 V, as shown in Figure 7.28 (note signal inversion, logical 1 being negative with respect to logical 0). Practical implementations use **+12 V** for **logical 0** and **−12 V** for **logical 1**.

RS-232 signals

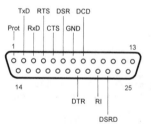

RS - 232C 25 way 'D' connector

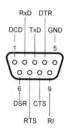

RS - 232C 9 way 'D' connector

Figure 7.29

The RS-232 standard specifies a 25 pin 'D' connector, although a 9 pin connector is often used nowadays with omission of certain signals (see Figure 7.29).

Most implementations of RS-232 do not make use of all of these signals, and some are specifically for use with modems. A simple RS-232 link may use only **three conductors**, **TXD**, **RXD** and **Com**. Descriptions of the most commonly used pins are as follows:

Protective ground: (Prot) This pin usually connects to the metal chassis of one device, and is connected via the cable to the corresponding connection on the other device.

Transmitted data: (TXD) This pin provides an output signal which transfers serial data from one RS-232 device to another, provided that the 'Clear to Send' (CTS) signal is active.

Received data: (RXD) This pin accepts serial input data which has been transmitted by some other RS-232 device.

Request to send: (RTS) This pin provides an output logic level which indicates that an RS-232 device is ready to transmit serial data. The state of this signal is tested by the receiving device so that it knows when to expect more data.

Clear to send: (CTS) This pin provides an output logic level which indicates that an RS-232 receiving device is ready to accept data via its 'Received data' input. It thus provides a signal to the transmitting device which inhibits the transfer of data until the receiving device is ready.

Data set ready: (DSR) This pin provides an output logic level which indicates that an RS-232 receiving device is ready for operation (i.e. switched on or operational).

Signal ground: (Com) This provides a common reference or ground for all input and output signals.

Data terminal ready: (DTR) This pin provides an output logic level which indicates that the RS-232 transmitting device is ready for operation (i.e. switched on or operational).

Although it is possible to transfer data using only **two lines** (signal plus ground), it is usual to make use of **handshake** signals such as **RTS** and **CTS** to enable the receiving device to control the flow of data. This arrangement avoids the receiving device from being presented with more data if it is still busy processing previous data, thus avoiding data loss.

RS-232 interfaces

Since RS-232 voltage levels are incompatible with TTL levels, some form of interface circuit is required to provide **voltage level conversion**. In addition, the logic levels must be **inverted**, as shown in Figure 7.30.

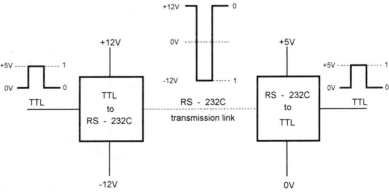

Figure 7.30

Interface circuits may be constructed from discrete components, or may make use of purpose-built interface devices such as the MAX232.

RS-232 to TTL

When receiving data, conversion of RS-232 signals to TTL voltage levels is relatively straightforward, requiring only limiting the swing of input signal between 0 V and +5 V, followed by signal inversion. Suitable circuits for this purpose, using both discrete and integrated solutions, are shown in Figures 7.31(a) and (b).

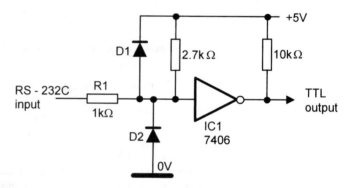

Figure 7.31(a)

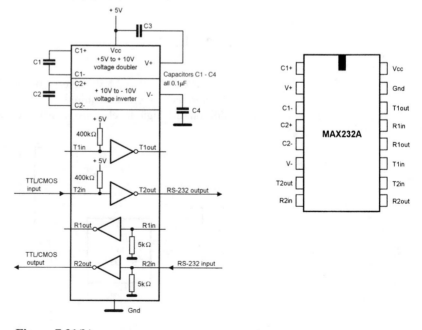

Figure 7.31(b)

The discrete component version shown in Figure 7.31(a) uses two **catching diodes**, D1 and D2, in conjunction with R1 to prevent the incoming RS-232 signal from exceeding the maximum permissible TTL input voltages. Signal inversion and buffering is provided by IC1, a 7406 inverter.

The integrated version is shown in Figure 7.31(b), and it is a very simple circuit, requiring few external components, since all conversion takes place within the IC which provides two identical RS-232 to TTL converters and two identical TTL to RS-232 converters.

TTL to RS-232

When transmitting data, the conversion of TTL signals to RS-232 levels is slightly more complicated, since in addition to inversion, it also requires the generation of **opposite polarity signals** at (usually) higher voltage levels. Typically the ±12 V levels of RS-232C are used. Suitable circuits for this purpose, using both discrete and integrated solutions, are shown in Figures 7.32 and 7.31(b).

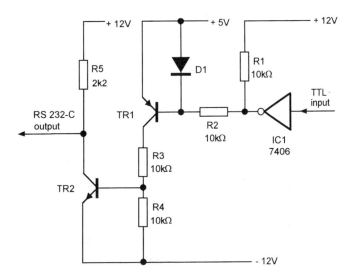

Figure 7.32

The discrete version shown in Figure 7.32 uses a 7406 'open-collector' IC to invert the TTL level source. The output signal may swing between 0 V and +12 V, and is applied to the base of TR1. Due to the presence of D1 and R2, the base of TR1 may swing approximately ±0.6 V with respect to the +5 V supply line. This is sufficient to cause the TR1 collector to switch between +5 V and −12 V. The voltage level available at the junction of R3/R4 biases the base of TR2 so that its collector potential switches between +12 V and −12 V, i.e. RS-232C levels.

The integrated version shown in Figure 7.31(b) uses the MAX232, since this device provides conversion in both directions. Again this is a very simple circuit, and two TTL to RS-232 converters are provided within the IC. One distinct advantage of the MAX232 compared with discrete circuits and the earlier 1488 IC is that it contains its own internal negative voltage generator, therefore the device may be operated from a single 5 V supply.

ASCII codes

Although not mandatory, serial data transfers usually take place using **ASCII** (*American Standard Code for Information Interchange*) codes. Up to 128 different characters and control codes are represented using 7-bit binary codes from 00_{16} to $7F_{16}$. An eighth bit may be used as a **parity bit** for **error checking** purposes, or for other special purposes, e.g. for a limited range of graphics codes. Standard ASCII codes are shown in Figure 7.33.

ASCII Character Codes	b6	0	0	0	0	1	1	1	1
	b5	0	0	1	1	0	0	1	1
	b4	0	1	0	1	0	1	0	1
b3 b2 b1 b0		0	1	2	3	4	5	6	7
0 0 0 0	0	NUL	DLE	SP	0	@	P	`	p
0 0 0 1	1	SOH	DC1	!	1	A	Q	a	q
0 0 1 0	2	STX	DC2	"	2	B	R	b	r
0 0 1 1	3	ETX	DC3	#	3	C	S	c	s
0 1 0 0	4	EOT	DC4	$	4	D	T	d	t
0 1 0 1	5	ENQ	NAK	%	5	E	U	e	u
0 1 1 0	6	ACK	SYN	&	6	F	V	f	v
0 1 1 1	7	BEL	ETB	'	7	G	W	g	w
1 0 0 0	8	BS	CAN	(	8	H	X	h	x
1 0 0 1	9	HT	EM	)	9	I	Y	i	y
1 0 1 0	A	LF	SUB	*	:	J	Z	j	z
1 0 1 1	B	VT	ESC	+	;	K	[	k	{
1 1 0 0	C	FF	FS	,	<	L	\	l		
1 1 0 1	D	CR	GS	-	=	M	]	m	}
1 1 1 0	E	SO	RS	.	>	N	^	n	~
1 1 1 1	F	SI	US	/	?	O	_	o	DEL

SP	Space		
DEL	Delete		
NUL	Null	DLE	Data link escape
SOH	Start of heading	DC1	Device control 1
STX	Start of text	DC2	Device control 2
ETX	End of text	DC3	Device control 3
EOT	End of transmission	DC4	Device control 4
ENQ	Enquiry	NAK	Negative acknowledge
ACK	Acknowledge	SYN	Synchronous idle
BEL	Bell, or alarm	ETB	End of transmission block
BS	Backspace	CAN	Cancel
HT	Horizontal tabulation	EM	End of medium
LF	Line feed	SUB	Substitute
VT	Vertical tabulation	ESC	Escape
FF	Form feed	FS	File separator
CR	Carriage return	GS	Group separator
SO	Shift out	RS	Record separator
SI	Shift in	US	Unit separator

Figure 7.33

Activity

1 Develop the **structure chart, pseudo-code** and **program code** to enable 2400 baud serial data (one start, eight data and two stop bits) to be read from one input line of a parallel data I/O port and to be converted into parallel form.

2 Connect a **2400 baud serial data** source to the input line (use a circuit similar to that shown in Figure 7.31(a) or (b) if using an RS-232 source such as serial output from another microcomputer).

3 Test your software for correct operation by displaying or storing the received data.

4 Write a report on your system, showing the **circuit diagram, software development** and **appraisal** of the performance of your system.

Digital to analogue conversion

A digital to analogue converter (DAC) is a device that converts a **multi-bit digital signal** at its input into an equivalent **analogue output signal,** i.e. it generates a different output potential for each binary input. The number of different output potentials therefore depends upon the number of digital inputs to the DAC. A digital to analogue converter is actually a multiplier circuit, and has a transfer function that may be expressed as:

$$\mathbf{f = a \times b} \qquad \textit{where} \ \ a = \textit{digital input,}$$
$$\textit{and} \quad b = \textit{analogue reference I or V}$$

The digital input 'a' is expressed as a fraction of the maximum value obtainable with the number of bits used. The general expression for the digital output is therefore:

$$\mathbf{V_{OUT} = (a/2^n) \times V_{REF}}$$

but since most DAC inputs are 8-bit, this is usually expressed as:

$$\mathbf{V_{OUT} = (a/256) \times V_{REF}}$$

For example, a **half-scale** digital input of 10000000_2 (128_{10}) produces a value for '**a**' of **128/256** or **0.5**. The maximum or **full-scale** input is $\mathbf{11111111_2}$ (255_{10}) when using eight bits, therefore the full-scale output is:

$$\mathbf{V_{OUT} = (255/256) \times V_{REF}}$$

i.e. for an 'n'-bit input, the full-scale is ($2^n - 1$) and the corresponding analogue output falls short of V_{REF} by an amount equivalent to that contributed by the least significant bit, as shown in Figure 7.34.

Most DAC converters make use of summing circuits, containing R-2R resistor ladder networks which ensure that each bit of the digital input signal contributes the correct amount towards the analogue output potential. The average user is, however, unlikely to be concerned with the internal construction of a DAC.

DAC characteristics

A large number of integrated circuit DACs are available from various manufacturers. When selecting a DAC for a particular application, reference should be made to the manufacturer's data sheets to determine its suitability. The following parameters and definitions may need to be considered.

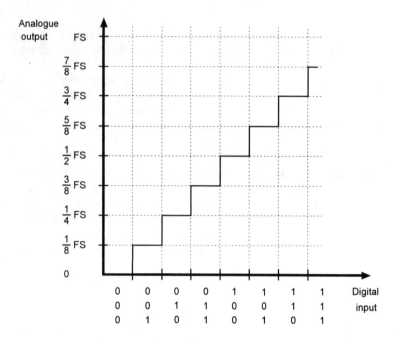

Figure 7.34

Resolution

The resolution of a DAC refers to the **number of input bits** it possesses, and indicates the smallest incremental change in analogue output voltage. For example, if a DAC has 'n' input bits, the number of increments is $2^n - 1$, and the magnitude of each increment is given by the expression:

$$ V = \frac{V_{REF}}{2^n - 1} $$

Monotonicity

When the input code to a DAC is increased in steps of **one LSB**, the analogue output voltage should also increase in steps of magnitude dependent upon the resolution of the converter. If the output always changes in this manner, the DAC is said to be **monotonic** since the output is a single valued function of the input, but if any step in this progression results in a decrease in DAC output voltage, the DAC is **non-monotonic** (see Figure 7.35).

The monotonicity of a DAC may be expressed in terms of the number of bits over which monotonicity is maintained.

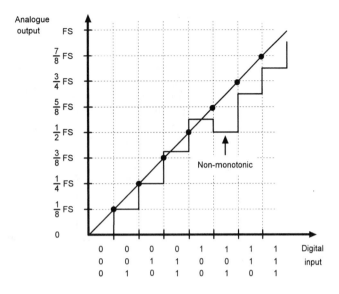

Figure 7.35

Offset (zero scale error)

Assuming unipolar operation and natural binary inputs, an input code of zero to a DAC should produce an analogue output of zero.

Due to imperfections in components and manufacturing techniques, a small offset may exist so that the transfer characteristic no longer passes through zero (see Figure 7.36).

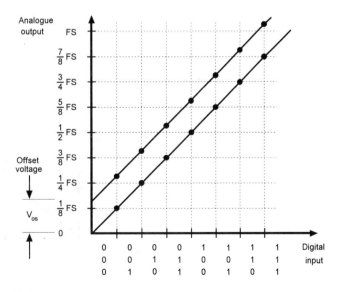

Figure 7.36

Gain

The gain of a DAC is an **analogue scale factor** that describes the relationship between the converter's full-scale output and its reference input (V_{REF}). The gain is usually adjusted by the user to *full-scale $x (1 - 2^{-n})$* with all input bits at logical 1. Ideally, the transfer characteristic then progresses linearly from zero to full-scale as its binary inputs sequence from minimum to maximum value. Imperfections in a DAC may cause a deviation from the ideal, as shown in Figures 7.37(a) and (b).

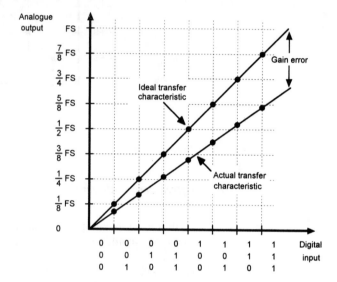

Figure 7.37(a)

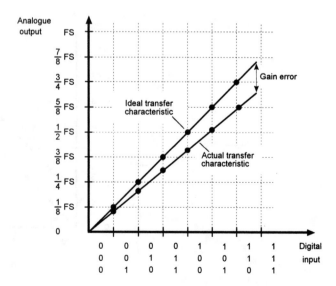

Figure 7.37(b)

The difference in slope between actual and ideal transfer characteristics is known as the **gain error**.

Linearity

Offset and gain errors may be 'trimmed' so that the end points of the characteristic lie on zero and full-scale, but this does not guarantee that intermediate points all lie on the ideal line. Linearity is a measure of how closely the analogue output characteristic of a DAC conforms to the ideal. This is usually quoted as a **linearity error** and is the deviation of the analogue output from an ideal straight line, expressed in % or **ppm** of the full-scale range or as a **fraction of one LSB** (see Figure 7.38).

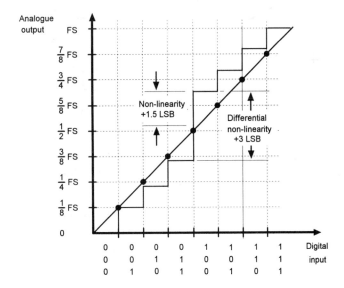

Figure 7.38

A linearity error within **±1/2 LSB** assures monotonic operation, although the converse is not true and a monotonic DAC may have large linearity errors.

Settling time

The settling time is the time taken for a DAC to settle to within **±1/2 LSB** of its final value after a transition in its input code. This is typically **100 ns**, but varies according to the number of bits that change. A single bit change, e.g. 00000000 to 00000001, gives the fastest figure while a transition which results in all bits changing, e.g. 01111111 to 10000000, gives the slowest settling time figure.

Bipolar operation

Previous theory has considered the output from a DAC as being single polarity (unipolar), i.e. the digital input produces an analogue output voltage between zero and some positive value. Some practical applications require a bipolar output voltage, i.e. **positive** and **negative** output voltage, and this may be achieved by the application of a negative offset of $V_{REF}/2$ to the analogue output voltage, as shown in Figures 7.39(a) and (b).

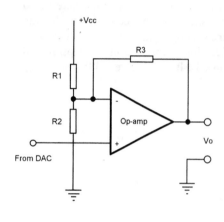

Figure 7.39(a)

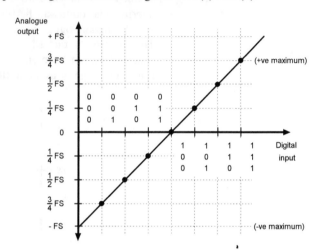

Figure 7.39(b)

Interfacing a DAC to a microcomputer

A DAC may be interfaced to a microcomputer in one of the following ways:

1 connected directly to the parallel outputs of a PPI device (port interface), or
2 connected to a microcomputer data bus (bus interface).

Some DACs do not have built-in latches and therefore are more readily interfaced to a parallel output port such as that provided by an 8255 PPI. One example of this type of DAC is the ZN426, and the block and pin out diagrams for this device are shown in Figure 7.40.

It can be seen that this DAC is not provided with any control inputs to enable direct connection to a microcomputer data bus, therefore it must be interfaced via a parallel output port, as shown in Figures 7.41(a) and (b). The circuit shown in Figure 7.41(a) does not require software configuring, and data is sent to the DAC by writing to I/O address 88h (this address may be changed by selecting different outputs from IC1 and IC2). If the 8255 PPI circuit shown in Figure 7.41(b) is used, this must first be configured as an output port before writing to the DAC.

Both of these circuits make use of a ZN424 buffer stage which enables both offset and gain to be adjusted

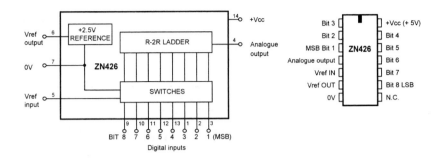

Figure 7.40

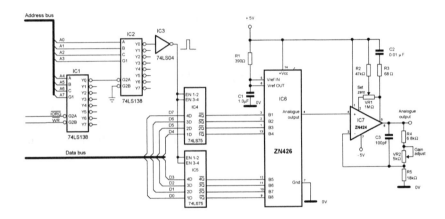

Figure 7.41(a)

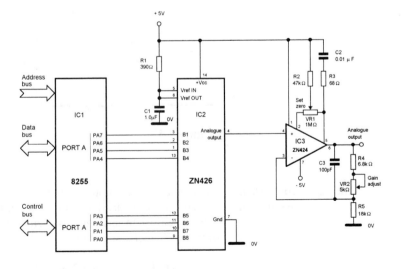

Figure 7.41(b)

Another type of DAC is the ZN428 which is provided with data latches, enabling it to be interfaced directly to a microcomputer data bus (see Figure 7.42).

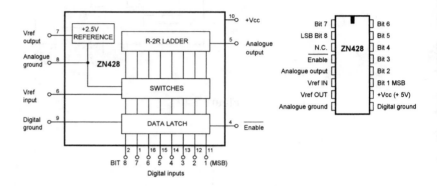

Figure 7.42

A negative going latch pulse is used to write data to the ZN428 DAC, and this pulse may be obtained from an address decoder, as shown in Figure 7.43. Data is latched into the DAC on the rising edge of the decoder pulse.

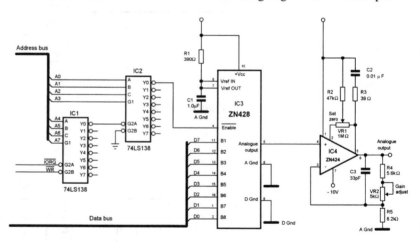

Figure 7.43

Analogue to digital conversion

Analogue to digital conversion is a process that converts an **analogue input** signal into an equivalent **multi-bit digital output** signal. This process is performed by an analogue to digital converter (ADC) which has an ideal transfer characteristic expressed as:

$$\text{Vfs}\left(\frac{\text{B1}}{2} + \frac{\text{B2}}{4} + \frac{\text{B3}}{8} + \text{... } \frac{\text{Bn}}{2^n}\right) = \text{Vin} \pm \tfrac{1}{2}\text{ LSB}$$

The transfer characteristic of an ideal ADC is shown in Figure 7.44, which for simplicity is restricted to three-bit digital output.

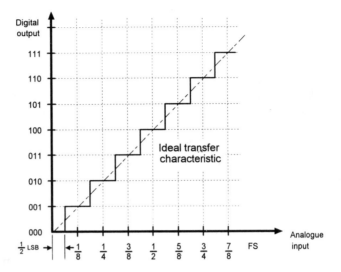

Figure 7.44

Using such an ADC, eight different digital outputs are available, 000_2 to 111_2 (0 to 7), which correspond to eight different analogue input voltage levels. In a practical situation, the analogue input seldom jumps from one value to the next but may vary continuously between 0 and full-scale. Therefore each digital output code represents a *range of* analogue input voltages, equivalent to that required to cause a change in output of one LSB. Zero of an ADC is usually adjusted so that changes in output code occur $\pm 1/2$ LSB either side of the actual analogue input corresponding to that code. For example, the output code 010_2 represents an analogue input of 1/4 Vfs, but the change from 001 to 010 occurs at 3/16 Vfs and the change from 010_2 to 011_2 occurs at 5/16 Vfs.

Various methods of performing analogue to digital conversion exist, most of which perform their task by comparing the analogue input with potentials generated from a digital source. The value of the digital source is varied until the potential it generates is equal to the analogue input and conversion is complete. This process may be carried out almost entirely in hardware, using a purpose-built ADC device. Alternatively, hardware consisting of DAC and voltage comparator circuits may be used, with software playing a major part in the conversion process.

Binary counter ADC

This is also known as a **ramp** type of converter, and consists of a **binary counter**, **DAC** and **voltage comparator** arranged as shown in Figure 7.45(a).

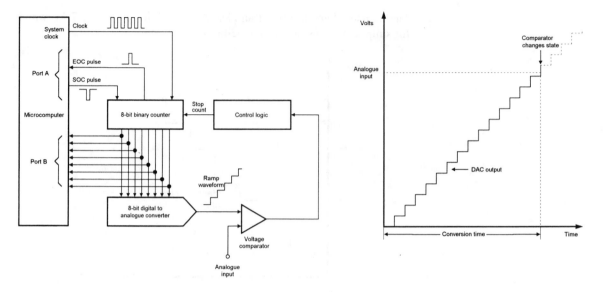

Figure 7.45(a)   **Figure 7.45(b)**

The counter is first reset to zero by a **start of conversion** (SOC) signal, and is then clocked so that its output increments in pure binary fashion which causes the DAC to generate a steadily rising (staircase) output voltage. The DAC output and the analogue input signal are both connected to the inputs of a voltage comparator circuit. If the analogue input voltage is greater than the DAC output voltage, the comparator output is logical 1 (high). This is the condition which prevails at the start of the count sequence. As the count progresses, the DAC output rises in staircase manner until its output voltage is equal to the analogue input voltage (see Figure 7.45(b)). At this point, the comparator output suddenly changes to a logical 0 (low) which causes the control logic circuit to terminate the counting sequence and latches the current count. An **end of conversion** (EOC) pulse is also generated. The residual count in the binary counter therefore represents the digital equivalent of the analogue input voltage.

The disadvantage of this conversion method is the time taken for the binary counter to reach its final value, which varies according to the analogue input being converted, and is relatively long when converting higher input levels. Although conversion time may not be a problem for many applications, nevertheless this technique is virtually never used in integrated ADC devices. In applications which allow the MPU to be wholly engaged in the conversion process, a counter may be maintained within the MPU by using one of its internal registers for this purpose. This allows the hardware to be simplified, and a circuit similar to that shown in Figure 7.46 may be used.

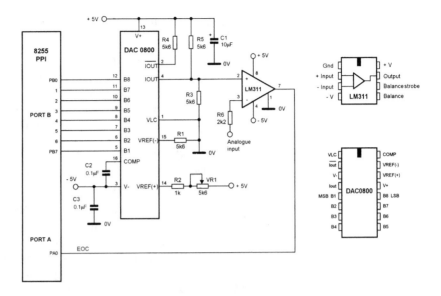

Figure 7.46

A structure chart for the software is shown in Figure 7.47.

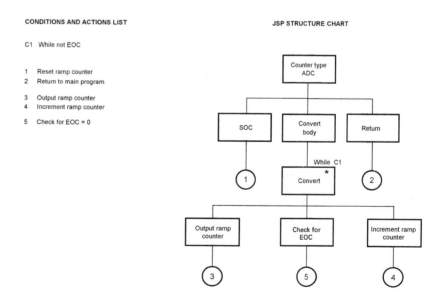

Figure 7.47

Pseudo-code for the successive approximation ADC subroutine is as follows:

```
COUNTER TYPE ADC seq
      DO    1                    {reset ramp count to zero}
      CONVERT BODY  iter  WHILE C1         {not EOC}
            DO    3              {output ramp counter to DAC}
            DO    5              {check for EOC (voltage comp. O/P = 0)}
            DO    4              {increment ramp counter}
      CONVERT BODY  end
      DO    2                    {return to main program}
COUNTER TYPE ADC end
```

The contents of a register within the MPU are incremented from zero, and the resulting binary count is sent out via the I/O port to the DAC. The comparator output (EOC) is tested at each stage of the count sequence, and the count is terminated when a logical 0 is detected.

The following listing shows a typical software subroutine for the Z80 MPU, but these are relatively simple to convert for other types of MPU using the pseudo-code provided.

```
           ; ********** Z80 counter type ADC ********
           ; PROGRAM 42
           ; performs counter/ramp type of conversion.
           ; Assumes PPI already configured.
           ; Invoked by calling 'adcon' and returns
           ; digital equivalent in register B.
           ; Uses circuit shown in Figure 7.46
           ; *****************************************
           ;
0010 =        ppi     EQU     10h          ; 8255 PPI base address
0010 =        status  EQU     ppi          ; EOC input
0011 =        dac     EQU     ppi+1        ; D to A converter
              ;
1900                  ORG     1900h        ; start of user RAM
              ;
              ; initialize ADC for conversion
              ;
1900 0E 11    adcon:  ld      c,dac        ; set up pointer to dac
1902 06 00            ld      b,0          ; zero ramp counter (SOC)
              ;
              ; output count and check for EOC
              ;
1904 ED 41    conv:   out     (c),b        ; generate ramp
1906 DB 10            in      a,(status)   ; read voltage comparator
1908 CB 47            bit     0,a          ; check for EOC
190A C8               ret     z            ; convert exit
              ;
              ; no EOC so increment counter for next step
              ;
190B 04               inc     b            ; ramp up one step
190C 18 F6            jr      conv         ; and continue
              ;
1900                  END     adcon
```

Activity

1 Connect an ADC circuit similar to that shown in Figure 7.46 to the I/O ports of a microcomputer.

2 Connect a suitable sensor to the ADC input, e.g. a potentiometer.

3 Using the structure chart given, write a suitable program to operate the circuit as a counter type of ADC, allowing a suitable delay between successive conversions. Arrange for the converted input to be displayed, e.g. LEDs connected to Port B outputs.

4 Check the operation of the system by running the program, varying the analogue input and verifying that the digital output varies in step with changes in analogue input potential.

5 Monitor the comparator output and determine the **range of conversion times** resulting from different analogue inputs.

6 Write a report on your findings.

Successive approximation ADC

The counter types of ADC described suffer from the fact that they are relatively slow in operation, and that the conversion time varies according to the analogue input voltage. The conversion time may be greatly reduced and made more consistent by using a conversion method known as **successive approximation**. This method involves setting each bit of the input to the DAC, in turn, starting with the most significant bit. After setting each bit, its effect is noted at the output from the voltage comparator, and if setting a particular bit results in the DAC output exceeding the analogue input voltage, the bit is reset again. This process is repeated for each bit in the register; therefore, for an 8-bit register only eight 'trial' outputs are required for any input voltage. This process is shown in Figures 7.48(a) and (b).

The hardware requirements for successive approximation depend upon whether the successive approximation register (SAR) is controlled by hardware or by software. If a successive approximation integrated ADC (IC) is selected, this may be interfaced to a microcomputer as shown in Figure 7.49.

The ADC0804 contains a complete successive approximation ADC. Conversion is started by writing to the ADC ($\overline{WR}$ low), and continues with

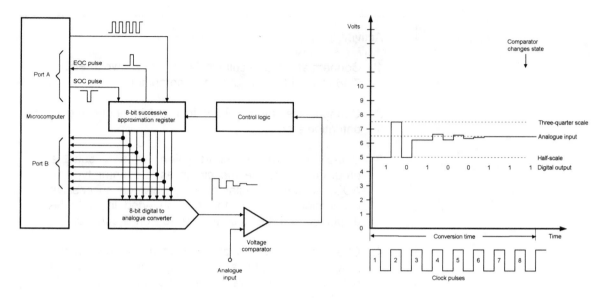

Figure 7.48(a)   **Figure 7.48(b)**

no further assistance from the MPU. Once conversion is finished, the ADC signals this by taking its $\overline{\text{INTR}}$ output low, and this may be used to interrupt the MPU, thereby causing it to read the digital data outputs DB0–DB7.

Software successive approximation may be used with the same hardware as that used for the software counter ADC, as shown in Figure 7.46. The successive approximation register is implemented with internal MPU registers.

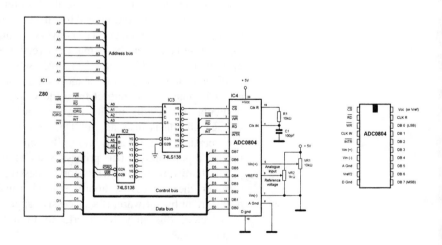

Figure 7.49

A structure chart for the successive approximation software is shown in Figure 7.50.

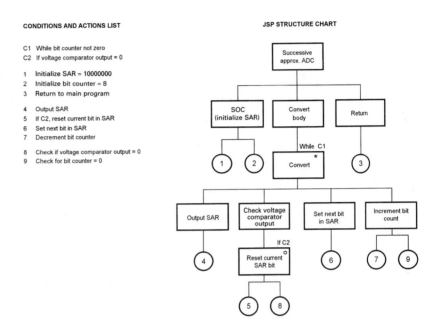

CONDITIONS AND ACTIONS LIST

C1 While bit counter not zero
C2 If voltage comparator output = 0

1 Initialize SAR = 10000000
2 Initialize bit counter = 8
3 Return to main program

4 Output SAR
5 If C2, reset current bit in SAR
6 Set next bit in SAR
7 Decrement bit counter

8 Check if voltage comparator output = 0
9 Check for bit counter = 0

JSP STRUCTURE CHART

Figure 7.50

Pseudo-code for the successive approximation ADC subroutine is as follows:

```
SUCCESSIVE APPROX. ADC    seq
    DO  1              {initialize SAR to 10000000}
    DO  2              {initialize bit count = 8}
    CONVERT BODY iter    WHILE C1 {bit counter not zero}
        CONVERT    seq
            DO  4      {output SAR to DAC}
            CHECK VOLT. COMP. O/P    sel  IF C2  {O/P = 0}
                DO    8    {check comparator output}
                DO    5    {reset current SAR bit}
            CHECK VOLT. COMP. O/P  end
            DO  6          {set next bit in SAR}
            DO  7          {decrement bit counter}
            DO  9          {check for bit counter = 0}
        CONVERT    end
    CONVERT BODY end
    DO  2                  {return to main program}
SUCCESSIVE APPROX. ADC    end
```

The following listing shows typical software subroutines for the Z80 MPU, but, using the pseudo-code provided, it is a relatively simple matter to develop subroutines for other types of MPU.

```
                              ; ********** Z80 successive approximation ADC *********
                              ; PROGRAM 43
                              ; performs successive approximation type of conversion.
                              ; Assumes PPI already configured.
                              ; Invoked by calling 'adcon' and returns
                              ; digital equivalent in register A.
                              ; Uses circuit shown in Figure 7.46
                              ; *****************************************
                              ;
0010 =          ppi          EQU    10h            ; 8255 PPI base address
0010 =          status       EQU    ppi            ; EOC input
0011 =          dac          EQU    ppi+1          ; D to A converter
                              ;
1900                         ORG    1900h          ; start of user RAM
                              ;
                              ; initialize ADC for conversion
                              ;
1900 0E 10      adcon:       ld     c,status       ; set up EOC pointer
1902 06 80                   ld     b,10000000b    ; B = SAR mask
1904 AF                      xor    a              ; A = SAR
                              ;
                              ; output SAR register and check status
                              ;
1905 B0         conv:        or     b              ; set SAR bit
1906 D3 11                   out    (dac),a        ; generate ramp
1908 ED 50                   in     d,(c)          ; read voltage comparator
190A CB 42                   bit    0,d            ; check current status
190C 20 01                   jr     nz,shift       ; too big?
                              ;
                              ; if too big, reset bit and try next bit for size
                              ;
190E A8                      xor    b              ; reset bit
190F CB 38      shift:       srl    b              ; shift SAR mask 1 bit
1911 30 F2                   jr     nc,conv        ; and continue
1913 C9                      ret
                              ;
1900                         END    adcon
```

Activity

1   Connect an ADC circuit similar to that shown in Figure 7.46 to the I/O ports of a microcomputer.

2   Connect a suitable sensor to the ADC input, e.g. a potentiometer.

3   Using the structure chart given, write a suitable program to operate the circuit as a **successive approximation** type of ADC, allowing a suitable delay between successive conversions. Arrange for the converted input to be displayed, e.g. LEDs connected to Port B outputs.

4   Check the operation of the system by running the program, varying the analogue input and verifying that the digital output varies in step with changes in analogue input potential.

5 Monitor the comparator output and determine the **conversion time**. Check that this time is constant and does not change with different analogue inputs.

6 Write a report on your findings.

ADC characteristics

A large number of monolithic ADCs are available from various manufacturers. When selecting an ADC for a particular application, reference should be made to the manufacturer's data sheets to determine its suitability. The following parameters and definitions may need to be considered.

Quantizing error

In a DAC, for each input code there is a fixed analogue output level, but for an ADC this is not the case. For each digital output level from an ADC there exists a range of analogue inputs equivalent to 1 LSB. Therefore it is not possible to determine the exact analogue input from its digital output code, there being the possibility of a **quantizing error** of **±1/2 LSB**. All ADCs therefore introduce a quantizing error whose magnitude depends upon the number of output bits of the ADC.

Missing codes

If the DAC used in an ADC is non-monotonic, then certain output codes cannot be generated. For example, consider the situation shown in Figure 7.51(a) which shows non-monotonicity on input code 101_2 (5).

If the analogue input is lower than the output generated by a DAC input code of 100_2 (4), then the counter stops before the code for 101_2 is reached. If the analogue input is greater than the DAC output for 100_2 (4) then it must also be greater than the output for 101_2 (5), therefore the output code for 101_2 (5) is never generated and is known as a **missing code**. ADC characteristics with a missing code are shown in Figure 7.51(b).

Zero transition

An ADC is usually adjusted so that changes from one output code to the next occur **±1/2 LSB either side** of the actual analogue input corresponding to

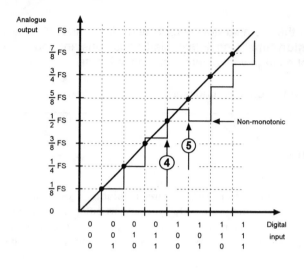

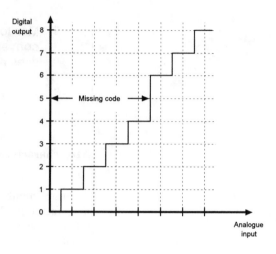

Figure 7.51(a)   **Figure 7.51(b)**

each code, i.e. the transition from 0 to 1 is offset by 1/2 LSB. The DAC used in an ADC circuit does not normally have such an offset, therefore the transition from 0 to 1 occurs at 1 LSB plus errors due to DAC zero error and comparator offset. The total of these is referred to as the **zero transition**.

Gain error

Similar to the gain error for a DAC, the gain error is the difference in slope between a line drawn from actual zero to full-scale and a line drawn through the ideal transition points.

Linearity

The linearity error of an ADC is the deviation of the actual characteristic from the ideal characteristic. It is specified as a percentage of full-scale or a fraction of an LSB, and if less than ±1/2 LSB, ensures that there are **no missing codes**.

Differential linearity

The differential linearity is a measure of the difference between an **actual** analogue increment required for a change from one output code to the next and the **ideal size**, $V_{FS}/2^n$. If the differential linearity is specified as +1/2 LSB, the step size from one state to the next may vary from half to one and a half times the ideal 1 LSB step.

Resolution

The resolution of an ADC is defined as the **number of output bits** it possesses. This determines the number of discrete output steps available but does not indicate the accuracy of an ADC.

Conversion time

The time taken for an ADC to complete the process of converting an analogue input into an equivalent digital output is known as the **conversion time**. For a successive approximation ADC, this time is constant and depends upon the number of bits and the clocking frequency. For a counter/comparator type of ADC, however, the conversion time varies according to the magnitude of the analogue input voltage, since this determines the number of clock pulses required before EOC is achieved. Generally the conversion time is quoted for a full-scale conversion and for 8-bit resolution is expressed as:

$$\textbf{Conversion time} \; = \frac{\textbf{256}}{\textbf{clock frequency (Hz)}} \; \textbf{s}$$

Interfacing protocols

It has been previously stated that timing of data transfers is an important factor for consideration when interfacing a microcomputer to its peripherals. Three basic protocols are commonly used when interfacing, depending upon the requirements of a system, and these are:

1 **Timed**
2 **Polled**
3 **Interrupt driven**

Timed I/O

Timed data transfers are commonly used when data transfers must take place at regular time intervals, or at predefined times. Examples of this type of transfer include **sequenced outputs** and **data logging**.

The use of software delay loops was considered in the previous chapter, and this technique may be used for timing of transfers in data sequencing or logging equipment. However, if long time intervals are required, software timing loops take up most of the processing time and the MPU is unable to perform any other tasks, e.g. *display refreshing*. This problem may be avoided by the use of a hardware timer which is controlled by the MPU, but which operates independently, generally by counting system clock pulses.

Polled I/O

Some peripherals require frequent attention, but at irregular time intervals and therefore cannot use timed I/O. In such cases software polling may be a suitable method. This is a process that is MPU initiated (via its program), in which the MPU repeatedly interrogates a peripheral device in order to determine whether it needs attention (see Figures 7.52(a) and (b)).

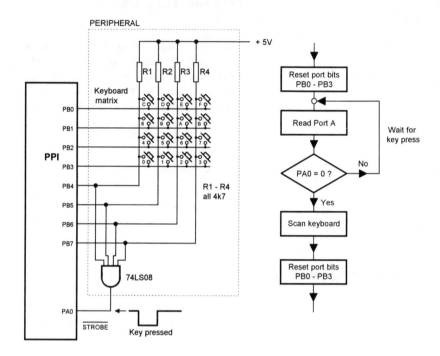

Figure 7.52(a)   **Figure 7.52(b)**

This diagram shows a typical polling situation where program behaviour depends upon an input from the user via a hex keyboard. The program must wait for a key press before it can continue. The hardware circuit shown in Figure 7.52(a) is arranged so that all inputs to the 74LS08 AND gate are at logical 1 until a key is pressed. The STROBE signal applied to PA0 falls to a logical 0 when any key is pressed, and the microcomputer must repeatedly poll this input to determine when this happens. Once a key press has been detected, a software scanning technique is used to determine which key has been pressed. A flowchart for this process is shown in Figure 7.52(b). It may take only a relatively short time period to service an individual peripheral, and a microcomputer may spend most of its time polling a peripheral that does not require servicing. It is therefore possible for a microcomputer to poll several peripherals, interrogating each in turn, and servicing only those that require attention (see Figures 7.53(a) and (b)).

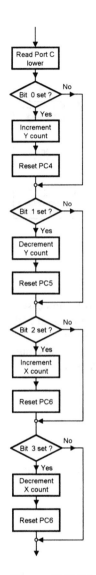

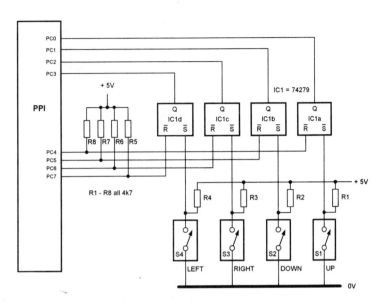

Figure 7.53(a)

Figure 7.53(b)

This system may be considered as four separate peripherals, S1 to S4, which may be separate push buttons or be in the form of a 'joystick'. Latches IC1(a) to (d) are used to remember that a switch has been operated in case the microcomputer is unable to deal with it immediately, and they also serve to debounce the switches. The polling routine shown in Figure 7.53(b) tests each of the inputs PC0 to PC3 in turn, and if active, takes appropriate action. Software is used to acknowledge polling of each peripheral and clears the appropriate latch.

An arrangement such as this assumes no peripheral requires immediate attention and can wait its turn in the polling cycle. The more peripherals there are to poll, the longer each one may have to wait for service, the worst-case situation being when a peripheral requests service just after it has been polled, with all following peripherals requesting service. If time is a critical factor, then the use of **interrupts** should be considered.

Interrupts

An interrupt is initiated externally by hardware connected to the **interrupt pin** of an MPU. Upon receipt of an interrupt request, an MPU completes its current instruction and then diverts from its main task to execute an **interrupt service routine (ISR)** thus providing an almost immediate response. An ISR is a subroutine that performs the task required by the interrupting peripheral.

This process is shown in Figure 7.54 which shows a switch connected to the interrupt input of the MPU to represent the interrupting peripheral.

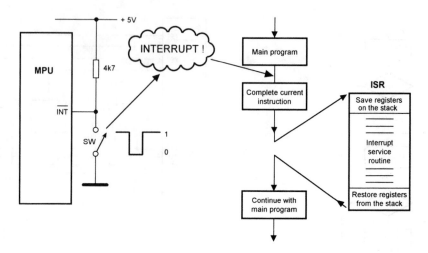

Figure 7.54

Since the action is initiated by the switch and not by the MPU, processing time is only taken when the switch requests it. Upon completion of an ISR, the main program continues as though it had never been interrupted.

Since the time at which an interrupt occurs is unpredictable, registers must be saved upon entering, and restored immediately prior to leaving the ISR, otherwise corruption of important registers may occur.

Interrupt response mechanism

The response of a microprocessor to an interrupt varies in detail according to the particular MPU being considered, but most perform the following operations:

(i) **Complete current instruction**

The INT pin of an MPU is checked internally during the final clock period of every instruction, thus ensuring that the MPU cannot be interrupted in the middle of an instruction.

(ii) **Check interrupt flag**

An interrupt may be ignored by an MPU if specifically **disabled**, or **masked**, by a software controlled **interrupt flag**. This prevents an MPU from being interrupted during critical parts of a program, e.g. while reading a file from disk, or receiving serial data. Once an MPU accepts an interrupt, it prevents further interrupts from being accepted by changing the state of its interrupt flag. This is done to prevent the ISR from being immediately (and repeatedly) interrupted if the INT pin is still active.

(iii) **Acknowledge interrupt**

Unless masked, an MPU acknowledges an interrupt request by changing the logic state on an **interrupt acknowledge** pin. This signal may be used by the interrupting device to place a byte onto the data bus to identify the starting address of the ISR (**vectored interrupt**).

(iv) **Save return address**

The return address (and usually the flags) are automatically pushed onto the stack. This ensures a correct return to the main program after execution of the ISR.

(v) **Load PC with ISR starting address**

The program counter is loaded (directly or indirectly) with the ISR starting address. Execution then continues from this address until an **interrupt return** instruction is encountered which is the last instruction of the ISR.

(vi) **Return from ISR, continue with main program**

When the return instruction is executed, registers saved on the stack by the MPU are reinstated with the values contained in them prior to responding to the interrupt. Therefore the main program continues as though it had not been interrupted.

Non-maskable interrupt (NMI)

Certain types of event (e.g. imminent power failure) require an MPU always to respond as rapidly as possible. Interrupts of this type must never be disabled and for this purpose an MPU is provided with an NMI input. Since this type of interrupt cannot be masked, the NMI input is **edge triggered** to prevent repeated interrupts during the time that the NMI input is in the active state. An MPU usually responds to an NMI request by jumping to a fixed address.

Maskable interrupts

An MPU may respond to an interrupt request on its INT (or INTR) input or it may ignore it, depending upon the state of the relevant interrupt flag. A summary of the interrupts available when using Z80, 8086 and 8031 MPUs is shown in Figures 7.55(a), (b) and (c).

Z80 Interrupt summary

$\overline{\text{NMI}}$		Negative edge triggered, execute ISR at starting address of 0066h.
$\overline{\text{INT}}$	Mode 0	Execute single instruction placed on data bus when interrupt is acknowledged (usually single byte RST instruction).
	Mode 1	Execute ISR at starting address of 0038h.
	Mode 2	Combine 8-bit I register with 8-bit vector placed on data bus by a peripheral when interrupt is acknowledged, use 16-bit bit address formed as pointer to access a table of ISR starting addresses.
IFF1 IFF2		Enable interrupts by setting IFF1 -- use EI instruction Disable interrupts by resetting IFF1 -- use DI instruction IFF2 stores state of IFF1 when NMI occurs and restores state of IFF1 upon completion of the ISR.

Figure 7.55(a)

It can be seen that interrupts may be enabled by setting an appropriate flag, and this may be achieved with the following instructions:

(a)	Z80	EI		*set interrupt enable flag (IFF1)*
(b)	8086	STI		*set interrupt enable flag*
(c)	8031	MOV	IE,#81h	*enable external interrupt 0*
	or	MOV	IE,#84h	*enable external interrupt 1*

After resetting an MPU, interrupts are **automatically disabled**. Once interrupts have been enabled, however, they may be disabled by executing the following instructions:

(a)	Z80	DI		*reset interrupt enable flag (IFF1)*
(b)	8086	CLI		*reset interrupt enable flag*
(c)	8031	MOV	IE,#0	*disable all interrupts*
	or	MOV	IE,#beh	*disable external interrupt 0 only*
	or	MOV	IE,#bbh	*disable external interrupt 1 only*

Interrupts are generally disabled during an ISR unless re-enabled by use of the appropriate enabling instruction.

Multiple interrupts

You may have noticed that an MPU has very few external interrupt pins, but practical systems often have a number of different interrupting peripherals.

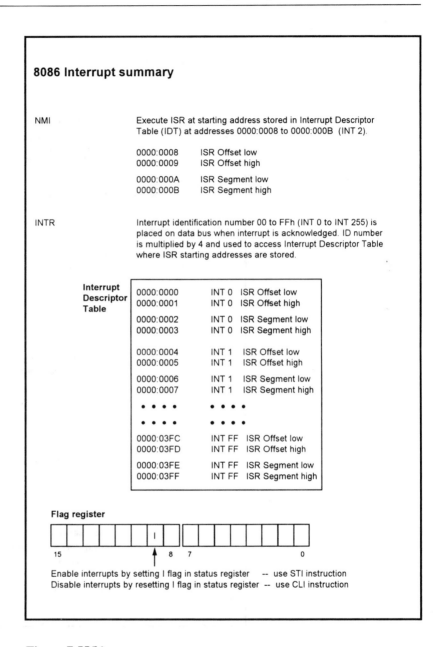

8086 Interrupt summary

NMI

Execute ISR at starting address stored in Interrupt Descriptor Table (IDT) at addresses 0000:0008 to 0000:000B (INT 2).

0000:0008	ISR Offset low
0000:0009	ISR Offset high
0000:000A	ISR Segment low
0000:000B	ISR Segment high

INTR

Interrupt identification number 00 to FFh (INT 0 to INT 255) is placed on data bus when interrupt is acknowledged. ID number is multiplied by 4 and used to access Interrupt Descriptor Table where ISR starting addresses are stored.

Interrupt Descriptor Table

0000:0000	INT 0	ISR Offset low
0000:0001	INT 0	ISR Offset high
0000:0002	INT 0	ISR Segment low
0000:0003	INT 0	ISR Segment high
0000:0004	INT 1	ISR Offset low
0000:0005	INT 1	ISR Offset high
0000:0006	INT 1	ISR Segment low
0000:0007	INT 1	ISR Segment high
• • • •	• • • •	
• • • •	• • • •	
0000:03FC	INT FF	ISR Offset low
0000:03FD	INT FF	ISR Offset high
0000:03FE	INT FF	ISR Segment low
0000:03FF	INT FF	ISR Segment high

Flag register

```
15                    I    8 7                    0
```

Enable interrupts by setting I flag in status register   -- use STI instruction
Disable interrupts by resetting I flag in status register -- use CLI instruction

Figure 7.55(b)

8031 Interrupt summary

$\overline{\text{INT0}}$	External interrupt, active low, execute ISR at starting address 0003
$\overline{\text{INT1}}$	External interrupt, active low, execute ISR at starting address 0013

TIMER 0	Internal interrupt when Timer 0 overflows, execute ISR at 000B
TIMER 1	Internal interrupt when Timer 1 overflows, execute ISR at 001B
TIMER 2	Internal interrupt when Timer 2 overflows, execute ISR at 002B (8032/52 only)

SERIAL PORT	Internal interrupt generated when serial RX buffer full or TX buffer empty, execute ISR at 002B (read SCON to determine whether RX or TX interrupt).

IE -- Interrupt Enable Register   (SFR address A8h)

EA	—	ET2	ES	ET1	EX1	ET0	EX0

EA	IE.7	Disables all interrupts if EA = 0 irrespective of states of individual enables.
—	IE.6	Not used
ET2	IE.5	Enable/Disable Timer 2 overflow interrupt (8032/52 only)
ES	IE.4	Enable/Disable serial port interrupt
ET1	IE.3	Enable/Disable Timer 1 overflow interrupt
EX1	IE.2	Enable/Disable external interrupt 1
ET0	IE.1	Enable/Disable Timer 0 overflow interrupt
EX0	IE.0	Enable/Disable external interrupt 0

IP -- Interrupt Priority Register   (SFR address B8h)

—	—	PT2	PS	PT1	PX1	PT0	PX0

—	IP.7	Not used
—	IP.6	Not used
PT2	IP.5	Define Timer 2 interrupt priority level (8032/52 only)
PS	IP.4	Define serial port interrupt priority level
PT1	IP.3	Define Timer 1 interrupt priority level
PX1	IP.2	Define external interrupt 1 priority level
PT0	IP.1	Define Timer 0 interrupt priority level
PX0	IP.0	Define external interrupt 0 priority level

Figure 7.55(c)

This problem can be solved in a number of different ways, for example all peripherals may share a single MPU interrupt pin, but when an interrupt request is received, the MPU polls each peripheral to determine the source of the interrupt (see Figure 7.56(a)).

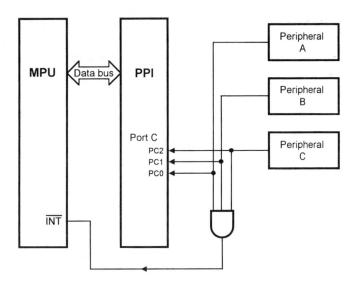

Figure 7.56(a)

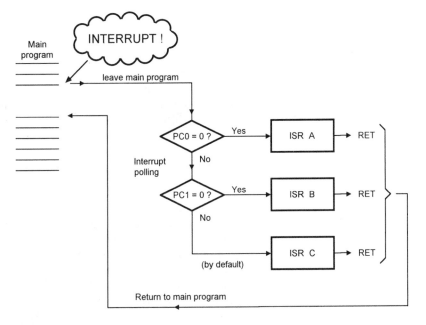

Figure 7.56(b)

Once the MPU recognizes an interrupt, it reads Port C of the PPI and tests bits 0 to 2 to identify the interrupting peripheral (see Figure 56(b)). Notice that testing of bit 2 may be omitted, since if neither peripheral A or peripheral B caused the interrupt, then it must have been peripheral C.

Alternatively, a peripheral may be able to place information on the data bus to enable the MPU to identify the interrupting source. This is only possible if the MPU responds with an interrupt acknowledge signal, and is known as a **vectored interrupt** (see Figure 7.57).

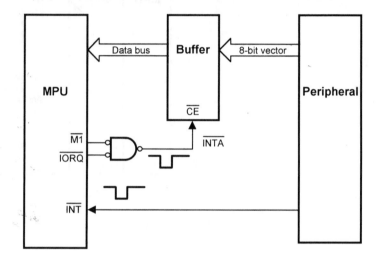

Figure 7.57

The buffer circuit input is an 8-bit vector that locates the ISR starting address. Once an interrupt request is recognized, the MPU generates an **interrupt acknowledge** signal that enables the buffer, thus placing the vector onto the data bus where it may be read by the MPU (the Z80 has no separate interrupt acknowledge pin, but instead activates two control signals, $\overline{M1}$ and $\overline{IORQ}$, which are not normally active at the same time).

Interrupt priority

A situation may occur in a multiple interrupt system when two peripherals interrupt simultaneously, therefore some form of priority is needed to ensure that the more important peripheral is serviced first. This may be managed with the interrupt system shown in Figure 7.56(a) by arranging to poll the peripherals in descending order of importance. When servicing a peripheral, the interrupt system must be disabled to prevent a low priority peripheral from interrupting the ISR of a more important peripheral, but this also unfortunately prevents a high priority peripheral from interrupting a low priority ISR.

True multi-level interrupts may be implemented by using a priority encoder which allows higher priority peripherals to interrupt an ISR, but masks out lower priority peripherals. A system of this type is shown in Figures 7.58(a) and (b)).

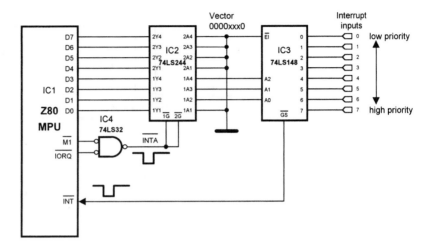

Figure 7.58(a)

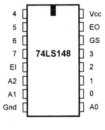

INPUTS									OUTPUTS				
EI	0	1	2	3	4	5	6	7	A2	A1	A0	GS	EO
1	X	X	X	X	X	X	X	X	1	1	1	1	1
0	1	1	1	1	1	1	1	1	1	1	1	1	0
0	X	X	X	X	X	X	X	0	0	0	0	0	1
0	X	X	X	X	X	X	0	1	0	0	1	0	1
0	X	X	X	X	X	0	1	1	0	1	0	0	1
0	X	X	X	X	0	1	1	1	0	1	1	0	1
0	X	X	X	0	1	1	1	1	1	0	0	0	1
0	X	X	0	1	1	1	1	1	1	0	1	0	1
0	X	0	1	1	1	1	1	1	1	1	0	0	1
0	0	1	1	1	1	1	1	1	1	1	1	0	1

Figure 7.58(b)

The priority encoder has eight interrupt inputs (0 to 7), and when one of these inputs is taken to logical 0, its corresponding three-bit binary code (inverted) appears on the A0 to A2 outputs, where it is used as part of an interrupt vector. The GS output is also activated (low) to request an interrupt. If an input of lower priority is activated, it is ignored, but one of higher priority will be accepted, and will interrupt the ISR for the lower priority request.

Because implementing such a system involves additional hardware logic circuits plus the necessary software, purpose-built integrated circuits are often used for this purpose. One such example is the **8259 Priority Interrupt Controller** shown in Figure 7.59.

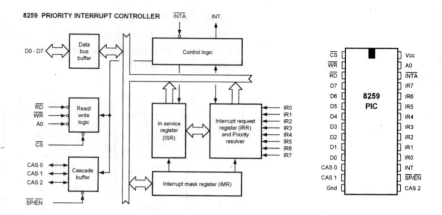

Figure 7.59

The programming of such a device is necessarily quite involved, therefore the following example may serve to indicate setting up interrupts using an 8259.

Example

A circuit of the type shown in Figure 7.60 is to be operated as a **decade counter** so that a level 0 interrupt occurs each time the push button is pressed and the count increments by one. The vector is stored in the IDT at **INT 20** index.

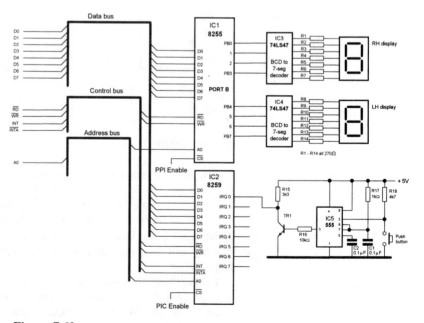

Figure 7.60

Software for this system is shown in the following 8086 assembly language program:

```
; ******** 8086 maskable interrupts ********
; PROGRAM 44
; uses 8259 PIC and 8255 PPI with 7-segment
; displays attached to Port B and operating
; as a decade counter, incrementing each
; time a push-button activated interrupt is
; issued by the 8259 (INT 0).
; Uses circuit shown in Figure 7.60
; *****************************************
;
                        .MODEL  SMALL
                        .STACK
                        ;
=  0010     ppi         EQU     10h         ; 8255 PPI base address
=  0011     display     EQU     ppi+1       ; 7-segment displays
=  0013     control     EQU     ppi+3       ; PPI control register
=  0090     iobyte      EQU     90h         ; I/O definition byte
=  0030     pic         EQU     30h         ; 8259 PIC base address
=  00FE     imask       EQU     11111110b   ; only IRQ 0 enabled
=  0017     icw1        EQU     00010111b   ; edge triggered (no icw3)
=  0020     icw2        EQU     20h         ; use INT 20
=  0003     icw4        EQU     3           ; 8086 auto, non-buffered
=  0080     int20       EQU     20h*4       ; INT 20 IDT entry

                        .CODE
0400                    ORG     0400h       ; start of user RAM
                        ;
                        ; initialize system
                        ;
0400 B0 90      start:  mov     al,iobyte   ; set up Port A as inputs
0402 E6 13              out     control,al  ; and Port B as outputs
0404 B8 0000            mov     ax,0        ; select IDT segment
0407 8E D8              mov     ds,ax
0409 C7 06 0080 042C R  mov     WORD PTR ds:int20,OFFSET isr ; initialize isr vector
040F C7 06 0082 ---- R  mov     WORD PTR ds:int20+2,SEG isr  ; in IDT (segment 0)
0415 B0 17              mov     al,icw1     ; interrupt control word 1
0417 E6 30              out     pic,al
0419 B0 20              mov     al,icw2     ; interrupt control word -2
041B E6 31              out     pic+1,al    ; (vector location/2)
041D B0 03              mov     al,icw4     ; interrupt control word 4
041F E6 31              out     pic+1,al
0421 B0 FE              mov     al,imask    ; activate IRQ 0
0423 E6 31              out     pic+1,al
0425 FB                 sti                 ; enable interrupts
0426 32 C0              xor     al,al       ; clear decade counter
0428 E6 11              out     display,al  ; and show zeros
                        ;
                        ; main program
                        ;
042A            main:   ;   *** main program instruction ***
                        ;   ***    should be entered here  ***
                        ;
042A EB FE              jmp     main        ; end loop

                        ;
                        ; interrupt service routine
                        ;
042C 04 01     isr:     add     al,1        ; increment BCD counter
042E 27                 daa
042F E6 11              out     display,al  ; and update displays
0431 FB                 sti                 ; re-arm interrupts
0432 CF                 iret                ; and return
                        ;
                        END     start
```

Many peripheral I/O devices such as the 8255 PPI or 8251 USART have output pins that may be used to request interrupts, generally when they are ready for the next data transfer operation.

When the 8255 PPI is operated in Mode 1 or Mode 2, then Port C bit 3 (PC3) acts as an interrupt request for Port A, and bit 0 (PC0) acts an interrupt request for Port B, and these bits may be connected to the 8259 IRQ inputs.

When the 8251 USART completes reception of a serial byte, its **RxRDY** output is activated, and if ready to transmit a serial byte, its **TxRDY** output is activated. Both of these outputs may be used to initiate interrupts by connecting these bits to the **8259 IRQ inputs**. These connections are shown in Figure 7.61.

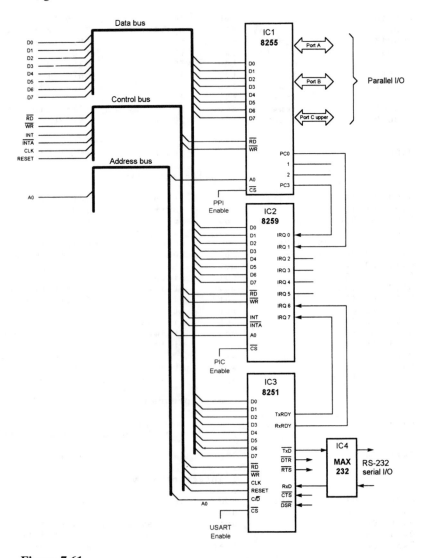

Figure 7.61

The 8031 MPU has built-in priorities such that each interrupting source may be assigned one of **two levels of priority**, determined by setting appropriate bits in the **IP register** (bit addressable). Interrupts may be requested from two external sources (**INT0** and **INT1**), **two timers** (TIMER 0 and TIMER 1), or the **serial I/O** system (TI and RI). A secondary system of priorities is also implemented for cases where any two or more of these sources with the same priority level simultaneously request an interrupt. The order is as follows:

Highest priority	IE0	(External interrupt 0)
	TF0	(Timer 0)
	IE1	(External interrupt 1)
	TF1	(Timer 1)
Lowest priority	RI or TI	(Serial port interrupt)

Test your knowledge 7.2

1 A parallel output port uses:
  A tri-state buffers to enable data transfers during an I/O read
  B data latches to enable data transfers during an I/O read
  C data latches to enable data transfers during an I/O write
  D tri-state buffers to enable data transfers during an I/O write

2 A parallel input port uses:
  A tri-state buffers to enable data transfers during an I/O read
  B data latches to enable data transfers during an I/O read
  C data latches to enable data transfers during an I/O write
  D tri-state buffers to enable data transfers during an I/O write

3 An interrupt is used to service a peripheral that requires:
  A urgent attention at fixed times
  B urgent attention at unpredictable times
  C non-urgent attention at fixed times
  D non-urgent attention at unpredictable times

4 The NMI input of a microprocessor:
  A is edge triggered but can be ignored
  B is edge triggered but cannot be ignored
  C is level triggered but can be ignored
  D is level triggered but cannot be ignored

5   The speed of serial data transfers is usually expressed in terms of its:
A   bit rate
B   baud rate
C   character rate
D   byte rate

6   The normal voltage levels for RS-232 serial data transfers are:
A   logical 0 = 0 V, logical 1 = +12 V
B   logical 0 = −12 V, logical 1 = +12 V
C   logical 0 = −12 V, logical 1 = 0 V
D   logical 0 = +12 V, logical 1 = −12 V

7   An ADC that does its conversion by applying a gradually increasing sequence of digital outputs to a voltage comparator is known as a:
A   ramp converter
B   successive approximation converter
C   flash converter
D   tracking converter

8   An ADC that does its conversion by setting and resetting each bit of a register in sequence is known as a:
A   flash converter
B   tracking converter
C   successive approximation converter
D   ramp converter

Activity

1   Study the interrupt facilities of any microcomputer available.

2   Set up hardware consisting of a microcomputer, a suitable output peripheral, e.g. 7-segment display or stepper motor, and a means of interrupting the MPU, e.g. a push-button switch.

3   Devise software for the system to cause the output peripheral to step each time an interrupt occurs.

4   Devise similar software to cause the output peripheral to step each time the push-button is operated by using a polling technique.

5   Write a report on your system, including circuit diagram and software details, and compare the interrupt and polling protocols.

Answers to test your knowledge questions

2.1   1 (B); 2 (D); 3 (C); 4 (A); 5 (A); 6 (C); 7 (D); 8 (C)
2.2   1 (C); 2 (D); 3 (A); 4 (D); 5 (B); 6 (A); 7 (D); 8 (C)
2.3   1 (A); 2 (D); 3 (A); 4 (B); 5 (D); 6 (D); 7 (B); 8 (A)

3.1   1 (A); 2 (D); 3 (B); 4 (B); 5 (B); 6 (D); 7 (C); 8 (D); 9 (B)
3.2   1 (C); 2 (D); 3 (B); 4 (D); 5 (A); 6 (C); 7 (C); 8 (B); 9 (D)

4.1   1 (D); 2 (B); 3 (D); 4 (C); 5 (D); 6 (C); 7 (D); 8 (A)
4.2   1 (D); 2 (B); 3 (A); 4 (D); 5 (A); 6 (D); 7 (D); 8 (A); 9 (A); 10 (C)

5.1   1 (C); 2 (B); 3 (A); 4 (D); 5 (C); 6 (D)

7.1   1 (D); 2 (C); 3 (D); 4 (B); 5 (D); 6 (C); 7 (A); 8 (D)
7.2   1 (C); 2 (A); 3 (B); 4 (B); 5 (B); 6 (B); 7 (A); 8 (C)

Index